Service Efficient Network Interconnection via Satellite

Service Efficient Network Interconnection via Satellite

EU Cost Action 253

Edited by

Y. Fun Hu
University of Bradford, UK

Gérard Maral
Ecole Nationale Supérieure des Télécommunications, France

Erina Ferro
CNUCE/C.N.R., Italy

JOHN WILEY & SONS, LTD

Other Wiley Editorial Offices

John Wiley & Sons, Inc., 605 Third Avenue,
New York, NY 10158-0012, USA

WILEY-VCH Verlag GmbH
Pappelallee 3, D-69469 Weinheim, Germany

John Wiley & Sons Australia Ltd, 33 Park Road, Milton,
Queensland 4064, Australia

John Wiley & Sons (Canada) Ltd, 22 Worcester Road
Rexdale, Ontario, M9W 1L1, Canada

John Wiley & Sons (Asia) Pte Ltd, 2 Clementi Loop #02-01,
Jin Xing Distripark, Singapore 129809

British Library Cataloguing in Publication Data

A catalogue record for this book is available from the British Library

ISBN 0471 48669 8

Typeset in Times by Deerpark Publishing Services Ltd, Shannon, Ireland.

Contents

Preface   ix

A Note from the COST253 Chairman   xi

Acronyms   xiii

Figures   xix

Tables   xxiii

Contributors   xxv

1 Introduction   1
1.1   Evolution of Satellite Communications   2
1.2   EU Initiatives in Satellite Communications   3
1.3   Operating Frequency   4
1.4   Technical Considerations   4
    1.4.1 Terminal   5
    1.4.2 Propagation   5
    1.4.3 Orbital Types   5
    1.4.4 On-Board Routing   6
    1.4.5 Traffic Management   6
1.5   Objectives and Activities of COST253   7
    1.5.1 Objectives   7
    1.5.2 Action Activities   7
1.6   Outline of Contents   10
    1.6.1 Chapter 2: Appropriate Traffic Generators for the Simulation of Services Supported
        by Non-GEO Constellations   11
    1.6.2 Chapter 3: Transmission Schemes   11
    1.6.3 Chapter 4: Networking   11
    1.6.4 Chapter 5: Evaluation Tools   12
References   12

2 Appropriate Traffic Generators for the Simulation of Services Supported by Non-GEO
   Constellation   15
2.1   Source Traffic Parameters and Descriptors   16
    2.1.1 Peak Cell Rate and Cell Delay Variation Tolerance   16
    2.1.2 Sustainable Cell Rate and Intrinsic Burst Tolerance   17
    2.1.3 Mean Burst Duration   18
    2.1.4 Burstiness   18
2.2   Quality of Service Parameters   18
2.3   ATM Service Categories   19

	2.3.1 *Constant Bit Rate (CBR) Service Category*	19
	2.3.2 *Variable Bit Rate (VBR) Service Category*	19
	2.3.3 *Available Bit Rate (ABR) Service Category*	20
	2.3.4 *Unspecified Bit Rate (UBR) Service Category*	20
	2.3.5 *Guaranteed Frame Rate (GFR) Service Category*	21
2.4	Statistical Behaviour of Traffic Sources	21
	2.4.1 *Voice*	21
	2.4.2 *Video*	22
	2.4.3 *Data*	23
	2.4.4 *Multimedia*	24
2.5	Influences on Traffic Characteristics	25
2.6	Source Models	26
	2.6.1 *Multi-State Markov Source Models*	26
	2.6.2 *Self Similar Models*	32
2.7	Geographic Traffic Models	33
	2.7.1 *Traffic Intensity – Forecasting Techniques*	33
References		34

3 Transmission Schemes | | **37**
3.1	Modulation Techniques	39
	3.1.1 *Overview of Modulation Schemes for Satellite Transmission*	39
	3.1.2 *Study of a Particular Modulation Scheme: Variable Rate N-MSK*	42
3.2	Coding Techniques	49
	3.2.1 *Overview of Channel Codes for Satellite Transmission*	49
	3.2.2 *Study of a Particular Channel Code – TPC*	54
3.3	Synchronisation	56
	3.3.1 *Overview of Required Synchronisation for Satellite Transmission*	56
	3.3.2 *Study of Döppler Frequency Shift Compensation for Mobile Satellites*	57
3.4	Catching Co-Channel Interference	70
	3.4.1 *Satellite System Model*	70
	3.4.2 *IPhP3 with Deterministic Marks*	71
	3.4.3 *Two First Moments of Cumulated Interference Power*	72
3.5	Chapter Summary and Perspectives	75
References		77

4 Networking | | **79**
4.1	LAN Interconnection	80
	4.1.1 *Introduction*	80
	4.1.2 *Satellite Network Architecture*	80
	4.1.3 *Terrestrial/Satellite Network Termination Module Characteristics*	82
	4.1.4 *Satellite Constellations*	83
	4.1.5 *Satellite Payload Architecture*	85
	4.1.6 *Satellite Payload Reference Functional Architecture*	87
	4.1.7 *ISLs*	91
4.2	Resource Control	91
	4.2.1 *Resource Allocation*	91
	4.2.2 *Call Set-Up and Routing*	94
4.3	Congestion Control	125
	4.3.1 *Overview*	125
	4.3.2 *The LEO Satellites Network Environment*	125
	4.3.3 *Network Scenario*	126
4.4	Multicast	127
	4.4.1 *The Broadband Constellations*	127
	4.4.2 *Multicast Technology*	130

4.5   Reliability   132
        4.5.1 Background   132
        4.5.2 The Modelling and Simulation Method   134
        4.5.3 Payload Reference Functional Architecture and Model   136
        4.5.4 Simulation of the Cycle-of-Life of a Constellation   138
4.6   Security   139
        4.6.1 Background   139
        4.6.2 Status of Current Research in Security in Communication Networks 140
        4.6.3 Security Services Implementation Issues   141
        4.6.4 IP Security–IPSEC   144
        4.6.5 Security for ATM over Satellite   145
4.7   Security Infrastructure   146
4.8   Conclusions   147
References   148

5 Evaluation Tools   **153**
5.1   An Overview of Network Simulators   154
5.2   LeoSim: A Simulator for Routing   155
        5.2.1 Network Model as Supported by LeoSim   156
        5.2.2 Introducing Versatility in the Network Model   158
        5.2.3 Implementation Issues   160
5.3   GaliLEO: A Framework for Joint Expertise   163
        5.3.1 Both Global and Limited Coverage   164
        5.3.2 The Architecture   164
        5.3.3 Assumptions and Definitions   165
        5.3.4 Logical Behaviour   166
        5.3.5 The Connection Set-up   170
        5.3.6 Traffic Over a Connection   171
        5.3.7 The Topology   172
        5.3.8 Routing   174
        5.3.9 Fault Management   175
        5.3.10 Some Implementation Aspects   175
        5.3.11 GaliLEO Methodology and Project Management   177
5.4   CONSIM™: A Complementary Tool for Reliability   178
        5.4.1 A Communicating Process Approach   181
        5.4.2 Customisation of the Model and the Control Scheme   182
        5.4.3 Development Status   182
5.5   AristoteLEO   183
        5.5.1 The User Interface   183
        5.5.2 The Architecture   184
5.6   SEESAWS: An Ambitious Concept   186
        5.6.1 Traffic Simulator   189
        5.6.2 Satellite Simulator   189
References   192

6 TCP/IP Over Satellite   **195**
6.1   Transmission Control Protocol   196
        6.1.1 Introduction   196
        6.1.2 Slow Start and Congestion Avoidance   196
        6.1.3 Fast Retransmission and Fast Recovery   197
        6.1.4 Selective Acknowledgement (SACK)   198
6.2   The Effects of Satellite Networks on TCP Performance   199
        6.2.1 Overview   199
        6.2.2 Large Bandwidth Delay Product   199

		6.2.3 *Long Delay Path*	200
		6.2.4 *Bit Error Rate versus Congestion*	201
		6.2.5 *Link Asymmetry*	201
		6.2.6 *Variable Round Trip Time*	203
		6.2.7 *Adaptation of Access and Satellite Backbone Networks*	203
		6.2.8 *Intermittent Connectivity*	205
		6.2.9 *Bandwidth Limitations*	205
6.3	Simulation Analysis		205
		6.3.1 *Network Architecture*	205
		6.3.2 *Simulation Results*	205
6.4	Fixed – Mobile Convergence		209
		6.4.1 *Introduction*	209
		6.4.2 *Mobile IP*	209
6.5	Further Research and Conclusions		212
References			212

Appendix A   **215**

Appendix B   **231**

Appendix C   **235**

Appendix D   **237**

Index   **239**

Preface

The use of satellites for the provision of services to remote and less-developed regions, where access by terrestrial fibre is not commercially attractive, is recognised as an efficient and economical means. The EU COST Action 253 was set up to study, define, model and simulate the inter-connection of local area networks through non-geostationary satellites for providing world-wide efficient interactive communication services.

This book summarises the results from activities carried out in COST253 during the past 4 years by a group of European experts in satellite communications. It covers a wide range of topics. The service-level aspects cover issues on traffic modelling, source modelling, service characterisation and geographic traffic modelling to incorporate the non-uniform traffic load in a Low-Earth-Orbit constellation. The radio transmission aspects cover modulation schemes, coding schemes, Doppler frequency shift correction and interference modelling. The network aspects incorporate results on routing strategies, reliability, discussions on multicasting and security. In addition, the challenges of Transmission Control Protocol/ Internet Protocol over satellite are also covered. A rather unusual chapter included in this book is the discussion on simulation tools and methodologies for satellite networks. It is hoped that this book can provide useful information for researchers in the field of satellite communications.

The advantages of COST (European Co-Operation in the field of Scientific and Technical Research) Actions are invaluable. With respective to the COST Action 253, a unique environment has been created to enablie several forms of collaborations:

- Sharing of expertise. The COST253 management committee consists of a mixture of industrial organisations and academic/research institutes; each contributes to different, yet complementary, expertise in the field of satellite communications. Sharing of knowledge and expertise is accomplished through technical meetings, presentations and workshops.
- Exchange of information and co-operative work through Short Term Scientific Missions (STSM)
- The STSM has proved to be one of the main advantages of co-operation. A network simulator, GALILEO, is developed through collaboration among different organisations under the support of STSMs, providing a constructive co-operation leading to the development of an open-source simulator platform.
- Facilitating the set-up of co-operative actions in other instances, including submission of proposals to other EU-supported frameworks.
- Facilitating opportunities to enhance knowledge of different technical aspects and in research interests/areas of different organisations. This Action offers the opportunity to pool resources

with other groups and individuals in the field of satellite communications for communal development of ideas, data and techniques.

- An opportunity for young scientists to present their work and get constructive feedback and comments, improving their skills with future benefit to industry, academia and research communities
- Providing, along with COST252 (which is concerned with the use of non-geostationary satellites for the provision of mobile and personal communication services), an international forum for satellite communications free from political and economical constraints, which is important for medium term research. The scientific interest and the potential future applications were the key factors in establishing COST253 Action.

The flexibility in the running of the Action enables in-depth technical contributions to be made without being bound to specific commitments. Additionally, co-operation and contacts have been established with university groups and industrial research laboratories through inviting their representatives in the WG activities in Workshops and Conferences. Presentations have been given by Matra Marconi Space, Alcatel Space Industries, ESA/ESTEC, Teledesic and Nortel resulting in fruitful exchanges of views. From the action, an initiative by some of the members was derived towards the issuing of a proposal named SEESAWS to the IST programme. This proposal was aimed at the development of a simulation environment for evaluating the performance of satellite networks. Co-ordination with other COST actions namely COST252 and COST255 has resulted in the set-up of a successful workshop in Toulouse, France (May, 1999).

A number of STSMs have contributed to exchanging information and know how among the research teams from several institutions: the Universidad Publica de Navarra, the University of Bradford, CNUCE/C.N.R., Ecole Nationale Supérieure des Télécommunications-Site de Toulouse, CSELT, DLR, Institut Jozef Stefan and Aristotle University of Thessaloniki.

To this end, we would like to acknowledge all the contributors and all the members of the COST253 Management Committee for their invaluable work, which serves as a basis for this book. We would also like to thank the chapter editors who have harmonised the submitted work from different workgroups to produce some meaningful outputs. Last but not least, we thank the European Commission for funding this COST activity.

Y. Fun Hu
(COST253 Budgetary Co-ordinator)
Gérard Maral
(COST253 Chairman)
Erina Ferro
(COST253 Vice Chair)

A Note from the COST253 Chairman

Action COST253 started in October 1996 and ended in September 2000. The first year was a bit slow in work due to the late funding from the European Commission, but this was compensated in subsequent years thanks to the enthusiasm and the dedication of all participants.

Action COST 253 has been a unique opportunity to co-ordinate activities of teams with common interest in satellite communications from many institutes and the industry all over Europe. As such, it has contributed to a better understanding of problems and solutions related to data transfer and delivery of interactive services over satellite links. The Action has mostly focused itself on non-geostationary systems as candidate systems for supporting such applications. This was in part due to the interest at the time for non-geostationary satellite constellations, and also because of the relatively less documented experience with such systems compared to geostationary ones.

The major results of the investigations made within Action COST 253 are reported in this book. Chapter 1 outlines the framework and lists the large variety of topics that have been addressed in the course of the Action. The subsequent chapters provide a deeper insight into these topics, from traffic generation to transmission schemes, networking, performance evaluation and service provision.

A significant output of the action is the generation of software for the evaluation of the performance of satellite systems for the provision of the target services. Here, I should mention the availability of the GaliLEO simulator which can be accessed on the web server http://galileo.tesa.prd.fr, and invite all interested readers to visit the site. It is my hope that the simulator will find many more supporters whose contributions will expand its capability in the future, to the benefit of the largest technical community.

I would like to thank all contributors to the present book, in particular, Doctor Fun Hu who acted as the main Editor and Doctor Erina Ferro as both Editor and Vice Chair of COST 253, for their effort in turning an overwhelming pile of 'temporary' documents into a well organised and consistent joint publication that reflects the outstanding work produced during the course of the Action. I am convinced that this book will prove invaluable to all engineers and students involved in the field.

Although the activities of Action COST253 have now come to an end, we all look forward to the continuation of the Action in the new Action COST 272 'Packet oriented service delivery via satellite'. We are delighted to welcome all interested parties both from academic and industry, and look forward to a new fruitful co-operation.

Professor Gérard Maral
Chairman, COST 253 Management Committee
Ecole Nationale Supérieure des Télécommunications
Site of Toulouse

Acronyms

ACTS	Advanced Communications Technologies and Services
ATM	Asynchronous Transfer Mode
ABR	Available Bit Rate
AAL	ATM Adaptation Layer
ARMA	Autoregressive Moving Average
ACI	Adjacent Channel Interference
AWGN	Additive White Gaussian Noise
AGC	Automatic Gain Control
ATM-AM	ATM Adaptation Module
API	Application Programme Interface
AH	Authentication Header
AS	Autonomous System
BCH	Bose-Chandhuri-Hocquenhem
BISDN	Broadband ISDN
BGP	Border Gateways Protocol
CA	Certification Authorities
CAC	Call Admission Control
CBC	Cipher Block Chaining
CBR	Constant Bit Rate
CBT	Core Based Trees
CDMA	Code Division Multiple Access
CDV	Cell Delay Variation
CDVT	Cell Delay Variation Tolerance
CER	Cell Error Ratio
CLP	Cell Loss Priority
CLR	Cell Loss Ratio
CMR	Cell Misinsertion Rate
CoA	Care of Address
COST	European Co-operation in the field of Scientific and Technical Research
CPM	Continuous Phase Modulation
CRL	Certificate Revocation List
CRP	Common Resource Pool
CTD	Cell Transfer Delay
CWND	Congestion Window
DBP	Delay Bandwidth Product
DCCH	Dedicated Control Channel

DDPSK	Double DPSK
DES	Digital Encryption Standard
DHCP	Dynamic Host Configuration Protocol
DMBS	Double Moveable Boundary Strategy
DPSK	Differential Phase Shift Keying
DQDB	Dual Queue Dual Bus
DQPSK	Differential QPSK
DS	Döppler Shift
DVMRP	Distance Vector Multicast Routing Protocol
EGP	Exterior Gateways Protocol
ESP	Encapsulated Security Payload
FD	Flow Deviation
FDDI	Fibre Distributed Digital Interface
FCG	Fundamental Coding Gain
FTP	File Transfer Protocol
FES	Fixed Earth Station
GBM	Generalised Bass Model
GCRA	Generic Cell Rate Algorithm
GDP	Gross Domestic Product
GEO	Geostationary Orbit
GFR	Guaranteed Frame Rate
GMDP	Geometrically Modulated Deterministic Process
GMSK	Gaussian Shaped MSK
GSM	Global System for Mobile Communications
GW	Gateway
HAAP	High Altitude Aeronautical Platform
HLR	Home Location Register
HPA	High Power Amplifier
IBP	Interrupted Bernoulli Process
IBT	Intrinsic Burst Tolerance
IDRP	Inter-Domain Routing Protocol
IETF	Internet Engineering Task Force
IGMP	Internet Group Management Protocol
IP	Internet Protocol
IP-AM	Internet Protocol Adaptation Module
IPhP3	Isotropic Phase-Type Planar Point Process
IPP	Interrupted Poisson Process
IPSEC	IP Security
IPX	Internet Packet Exchange
ISDN	Integrated Services Digital Network
ISI	Inter-Symbol Interference
IS-IS	Intermediate System to Intermediate System
ISL	Inter-Satellite Links
IST	Information Society Technologies
ITU	International Telecommunications Union
LAC	LAN ATM Converter

LAN	Local Area Network
LC	Link Cost
LEO	Low Earth Orbit
LLA	Log-Likelihood Algorithm
LLC	Logical Link Control
LPF	Low Pass Filter
LRC	Raised Cosine with Pulse Length L
LREC	Rectangular Frequency Pulse of Length L
LSRC	Spectral Raised Cosine of Length L
MAC	Medium Access Control
MAC	Mutual Authentication Code (as in security)
MAP	Markovian Arrival Process
MBS	Maximum Burst Size
MCR	Minimum Cell Rate
MEO	Medium Earth Orbit
MLSF	Maximum Likelihood Sequence Estimation
MMDP	Markov Modulated Deterministic Process
MOSPF	Multicast Open Shortest Path First
MPSK	Multilevel PSK
MQAM	Multilevel QAM
MSC	Mobile Switch Centre
MSK	Minimum Shift Keying
NCC	Network Control Centre
OBP	On-Board Processing
OBS	On-Board Switching
OSI	Open System Interconnection
OAM	Operation and Maintenance
OSPF	Open Shortest Path First
PAWS	Protection Against Wrapped Sequence Numbers
PCC	Parallel Concatenated Convolution
PCR	Peak Cell Rate
PDU	Packet Data Units
PIM-SM	Protocol Independent Multicast–Sparse Mode
PNNI	Private Network to Network Interface
POTS	Plain Old Telephony Service
PSK	Phase Shift Keying
QAM	Quadrature Amplitude Modulation
QDDPSK	Quadrature DDPSK
QDPSK	Quadrature DPSK
QoS	Quality of Service
QPSK	Quaternary Phase Shift Keying
RACE	Research and Technology Development in Advanced Communication Technologies in Europe
RBD	Reliability Block Diagram
RFC	Request for Comments
RIP	Routing Information Protocol

RM	Resource Management
RPL	Radio Physical Layer
RSSI	Received Signal Strength Indicator
RTT	Round Trip Time
RTO	Retransmission Time Out
SA	Security Association
SACK	Selective Acknowledgement
SAR	Segmentation and Reassembly
SCC	Serial Concatenated Convolution
SCR	Sustainable Cell Rate
SDH	Synchronous Digital Hierarchy
SECBR	Severely Errored Cell Block Ratio
SIR	Signal-to-Interference Ratio
SNTM	Satellite Network Termination Module
S-PCS	Satellite Personal Communications Services
SPD	Spectral Power Density
SPI	Security Parameter Index
SPP	Switched Poisson Process
SNR	Signal-to-Noise Ratio
SRES	Signal Response
SS-TDMA	Satellite Switch TDMA
STM	Synchronous Transfer Mode
TC	Transmission Convergence
TCH	Traffic Channel
TCM	Trellis Coded Modulation
TCP	Transmission Control Protocol
TFM	Tamed Frequency Modulation
TIA	Telecommunications Industries of America
TNL	Terrestrial Network Link
TNTM	Terrestrial Network Termination Module
TOS	Type of Service
TPC	Turbo Decoding of Product Codes
TTL	Time To Live
TWF	Traffic Weight Factor
UBR	Unspecified Bit Rate
UDL	Up/Down Link
UMTS	Universal Mobile Telecommunications Systems
UNI	User Network Interface
UPC	Usage Parameter Control
UT	User Terminal
VBR	Variable Bit Rate
VC	Virtual Circuits
VCI	Virtual Channel Identifier
VCO	Voltage Controlled Oscillation
VLR	Visitor Location Register
VP	Virtual Path

VPC	Virtual Path Connection
VPI	Virtual Path Identifier
VSAT	Very Small Aperture Terminal
VS/VD	Virtual Source/Virtual Destination
WAN	Wide Area Network
WML	Wireless Application Protocol Mark-up Language

Figures

Figure 1.1 Interconnections between different working groups   9
Figure 2.1 Influences on different protocol layers   25
Figure 2.2 Two-state Markovian representation of an ATM source   27
Figure 2.3 The on–off source model   28
Figure 2.4 Two-dimensional birth–death process   30
Figure 2.5 State-transmission-rate diagram for aggregate source model   30
Figure 2.6 Two-state MMPP   31
Figure 3.1 Ideal QPSK spectrum and spectral regrowth of a QPSK signal with 1 dB back-off
            and a roll-off $\alpha = 0.35$ 39
Figure 3.2 Typical 16-QAM constellation at non-linear channel output with AM/AM and AM/
            PM effects (without noise). The clustering of the constellation points is due to
            interaction between non-linearities and the memory of the Nyquist filtering   41
Figure 3.3 Phase tree of a CPM with $M = 2$, $L = 3$ and $h = 2/3$ [3]. The pulse is an RC   42
Figure 3.4 N-MSK signals scattering diagrams   43
Figure 3.5 N-MSK signals power spectra, linear amplifier   44
Figure 3.6 N-MSK signals power spectra, non-linear high power amplifier with operation point
            -3 dB   44
Figure 3.7 Simulation model of variable rate communication system   45
Figure 3.8 Bit error rate curves for N-MSK signals   47
Figure 3.9 Eye closure influence on BER for N-MSK signals   47
Figure 3.10 Average number of transmitted bits per symbol, if the receiver generated noise is
            12.5 dB lower than the signal average power   48
Figure 3.11 Average number of transmitted bits per symbol, if the receiver generated noise is
            10.0 dB lower than the signal average power   48
Figure 3.12 Usual organisation of a linear block code based on GF(2^q). If the code is binary
            $q = 2$ 50
Figure 3.13 Trellis diagram of a rate 1/2 four-state convolution code. The input bit determines
            the branch taken from a particular state at a certain moment. The solid lines represent
            the possible codewords of this code. Each branch represents two output bits   51
Figure 3.14 Functional representation of concatenated coding with deinterleaving for spreading
            of error bursts present at the Viterbi decoder output   51
Figure 3.15 Two-level set partitioning of a 16-QAM. Two coded output bits of a 1/2 convolu-
            tional encoder designate one of the subsets, while two uncoded bits designate one of
            the constellation points within the subset. The spectral efficiency is 3 bits/T_s 52
Figure 3.16 Typical example of a Turbo encoder/decoder   53
Figure 3.17 Two-dimensional product code matrix   55
Figure 3.18 BER vs. E_b/N_0 in dB for uncoded QPSK and Turbo-decoded square product codes
            based on BCH codes   57
Figure 3.19 Döppler shift variation as a function of time   58
Figure 3.20 Differential detection principle   60
Figure 3.21 Döppler shift correcting receiver   61
Figure 3.22 Principle of DDPSK transmitter and receiver   63
Figure 3.23 RMS phase vs. SNR classical feedback loop   65

Figure 3.24 BER vs. E_b/N_0 for different Döppler shifts with QDPSK modulation   66
Figure 3.25 RMS phase vs. SNR for QDPSK with DS = $1/10T$, $B = 2/T$ and $K\chi = 40$ 68
Figure 3.26 RMS vs. SNR for QDDPSK with $B = (1 + \alpha)/T$, $\alpha = 0.3$ and DS = $1/16T$ 69
Figure 3.27 Spectral efficiency as a function of the SNR for a BER = 10^{-5}. The Shannon limit
and some unmodulated linear PSK and QAM schemes are shown, as well as exam-
ples of coded schemes. The concept of Fundamental Coding Gain (FCG) is also
shown   76
Figure 4.1 Satellite network architecture   81
Figure 4.2 ATM Protocol layer stack for OBS satellite   82
Figure 4.3 Block diagram of ground station transmit functionalities   82
Figure 4.4 Functional architecture of a general payload   90
Figure 4.5 Configuration of user access mode   92
Figure 4.6 DBMS frame organisation   93
Figure 4.7 Snapshot of Odyssey-like constellation with ISLs   97
Figure 4.8 Snapshot of Celestri-like constellation with ISLs   97
Figure 4.9 Daily profile of user activity   99
Figure 4.10 Average packet delay in the Celestri-like network for asymmetric (RF = 0)
and symmetric (RF = 1) traffic load   102
Figure 4.11 Average number of hops in the Celestri-like network for asymmetric (RF = 0)
and symmetric (RF = 1) traffic load   102
Figure 4.12 Mean delay of received packets on satellites in one orbit period for asymmetric
(RF = 0) and symmetric (RF = 1) traffic load   103
Figure 4.13 Satellite with the highest normalised link load for TWF = 0 and RF = 1   104
Figure 4.14 Satellite with the highest normalised link load for TWF = 1 and RF = 1   104
Figure 4.15 Satellite with the highest normalised link load for TWF = 10 and RF = 1   105
Figure 4.16 Normalised total traffic load for (a) Odyssey-like constellation and (b) Celestri-like
constellation   106
Figure 4.17 Average delay in Odyssey-like network for different values of reply factor   107
Figure 4.18 Average delay in Celestri-like network for different values of reply factor   108
Figure 4.19 Average number of hops in Odyssey-like network for different values of reply factor   109
Figure 4.20 Average number of hops in Celestri-like network for different values of reply factor   110
Figure 4.21 FD convergence for symmetric network load   112
Figure 4.22 FD Convergence according to N_p for different selection criteria   113
Figure 4.23 FD, Dijkstra and adaptive Dijkstra according to different amounts of balanced
network load   113
Figure 4.24 FD, Dijkstra and adaptive Dijkstra for the case of unbalanced network load accord-
ing to variation with mean network load   114
Figure 4.25 FD, Dijkstra and adaptive Dijkstra for balanced network load and non-uniform
distribution of Earth stations   114
Figure 4.26 Real-time simulation for the Poisson case for the FD, Dijkstra and adaptive Dijkstra
with link flow algorithms with mean input load 0.4 packets/timeslot   115
Figure 4.27 Real-time simulation for the Poisson case for the FD, Dijkstra and adaptive Dijkstra
Algorithms with mean input load 0.7 packets/timeslot   115
Figure 4.28 Real-time simulation for the self-similar traffic for the Dijkstra, FD and adaptive
Dijkstra algorithms with mean input load 0.4 packets/timeslot   116
Figure 4.29 Real-time simulation for the self-similar case for the FD and adaptive Dijkstra
algorithms with mean input load 0.7 packets/timeslot   117
Figure 4.30 Terrestrial, UDL and inter-satellite link segments   117
Figure 4.31 Comparison of the route diversity for Iridium and M-Star2 ($\tau = 50\%$)   119
Figure 4.32 Connection blocking probability for static and adaptive routing   120
Figure 4.33 Connection blocking probability for Periodic and Triggered distribution schemes at
different periods and trigger thresholds (Gateway traffic intensity = 4 Erlangs)   122
Figure 4.34 Connection blocking probability for the pre-computed scheme at different pre-
computation periods (GW trafic intensity = 4 Erlangs). The horizontal line with
circle shaped points corresponds to on-demand blocking probablity   123

Figure 4.35 Teledesic satellites and ground coverage rendered by SaVi [57]   128
Figure 4.36 SkyBridge satellites and ground coverage rendered by SaVi [57]   129
Figure 4.37 Celestri LEO satellites and ground coverage rendered by SaVi [57]   129
Figure 4.38 Input and output parameters of the methodology   135
Figure 4.39 RBD model of the operational satellite   137
Figure 4.40 RBD model of the spare satellite   138
Figure 4.41 Reliability function for the operational and the spare satellite models 139
Figure 4.42 Probability Pr{Satellite Failed, t}   139
Figure 4.43 Probability Pr{No Service, t}   140
Figure 4.44 ATM protocol reference model   143
Figure 4.45 ESP of INSPEC   145
Figure 5.1   LeoSim different models   157
Figure 5.2   ISL routing components built based on generic ISL component   159
Figure 5.3   The model of the Earth   164
Figure 5.4   Main components of the GaliLEO architecture   165
Figure 5.5   Logical diagram of the architectural core of the simulator   167
Figure 5.6   The architectural core of the simulator from an implementation point of view 169
Figure 5.7   The topology module   172
Figure 5.8   General architecture of the CONSIM™ environment   179
Figure 5.9   CONSIM™ model and core simulator architecture   180
Figure 5.10 Processes of the space segment O&M model   181
Figure 5.11 AristoteLEO   185
Figure 5.12 Geographical module   185
Figure 5.13 Controller   186
Figure 5.14 Traffic engine   187
Figure 5.15 SEESAWS functional architecture   188
Figure 6.1   Slow start process   197
Figure 6.2   TCP SACK option format   199
Figure 6.3   TCP connection data transfer phase   200
Figure 6.4   TCP asymmetry   202
Figure 6.5   Split segment network architecture   204
Figure 6.6   Split TCP connections   204
Figure 6.7   Simulation scenario   206
Figure 6.8   Effects of delay to the TCP throughput efficiency   207
Figure 6.9   Effect of channel BER level   208
Figure 6.10 Transfer time improvements   208
Figure 6.11 Mobile IP datagram flow   210

Tables

Table 2.1   List of ATM service traffic categories parameters   20
Table 2.2   Provisional QoS network performance objectives   21
Table 3.1   Burst structure for variable rate communication system   46
Table 3.2   Squared product code adaptation to packet lengths   56
Table 3.3   Number of symbols needed for acquisition as a function of the LPF parameter χ 67
Table 4.1   Comparison of various switching techniques   88
Table 4.2   Parameters of the selected satellite systems   96
Table 4.3   Percentage of total traffic flow between source and destination region   98
Table 4.4   Simulation parameters for uniform traffic   101
Table 4.5   Queuing delay statistics for all ISLs   103
Table 4.6   The highest normalised link load for different values of TWF in case of symmetric traffic load (RF = 1)   105
Table 4.7   Simulation parameters   106
Table 4.8   Satellite constellation parameters   108
Table 4.9   Location and magnitude of the Earth stations   111
Table 4.10   Traffic levels   111
Table 4.11   Constellation characteristics   119
Table 4.12   Comparison of the two causes for blocking according to their percentage of occurrence with adaptive routing (Iridium case)   121
Table 4.13   Number of state information broadcasts per sample period of 1000 s   122
Table 5.1   Example of possible studies using LeoSim   159
Table 6.1   TCP channel utilisation   200

Contributors

Marco Annoni
Telecom Italia Laboratory
Italy

Javier Aracil
Universidad Publica de Navarra
Spain

Simone Bizzarri
Telecom Italia Laboratory
Italy

Pauline M.L. Chan
University of Bradford
UK

Haitham Cruickshank
University of Surrey
UK

Fairouz Dabbarh
Universite Libre de Bruxelles
Belgium

Erina Ferro
CNUCE/C.N.R
Italy

Laurent Franck
Ecole Nationale Supérieure des Télécommunications
France

Ioannis Gragopoulos
Aristotle University of Thessaloniki
Greece

Y. Fun Hu
University of Bradford
UK

Tomaz Javornik
Institut Jozef Stefan
Slovenia

Gorazd Kandus
Institut Jozef Stefan
Slovenia

Gérard Maral
Ecole Nationale Supérieure des Télécommunications
France

Mihael Mohorcic
Institut Jozef Stefan
Slovenia

Tolga Ors
University of Surrey
UK

Evangelos Papapetrou
Aristotle University of Thessaloniki
Greece

F. Niovi Pavlidou
Aristotle University of Thessaloniki
Greece

Vendela Maria Paxal
Nera Wireless Broadband Access AS
Norway
(formerly of Telenor R&D, Norway)

Francesco Potorti
CNUCE/C.N.R
Italy

Marie-Ange Remiche
Universite Libre de Bruxelles
Belgium

Ray Sheriff
University of Bradford
UK

Zilli Sun
University of Surrey
UK

Ales Svigelj
Institut Jozef Stefan
Slovenia

Denis Trcek
Institut Jozef Stefan
Slovenia

Lloyd Wood
University of Surrey
UK

1

Introduction

Distributed data processing has developed rapidly over the last decades due to the dramatic increase of processing power and advances in microelectronics and communications. The establishment and acceptance of standards led to a widespread use of local computer networks capable of supporting speeds ranging from several Mbps to hundreds of Mbps. Nowadays local area networks (LAN) can be found in nearly every office environment and even in small entities of industry and R&D organisations. Furthermore, user requirements and service demands for multimedia (voice, video and data) communications over long distances are growing rapidly as a result of the increasing internationalisation of work.

While networks can be set up easily and at comparatively low cost locally, long-distance communications at speeds relative to the LANs is still a problem. In highly developed regions (West Europe, US and Japan), digital networks based on optical fibres joining major cities have been or are being developed – offering very high transmission rates to meet bandwidth demands. However, it leaves large rural and developing regions (e.g. Eastern Europe, South America, parts of Asia and Africa) without a proper infrastructure for advanced data communications.

Communication satellites play an important role in the rapid development of a global communication infrastructure due to their wide geographical coverage, quick and cost-effective deployment and configuration flexibility. They provide seamless integration of applications and services, which have traditionally been available via terrestrial networks. In addition, they make reachable those remote and less developed regions where access by terrestrial fibres is not commercially attractive, due to the high installation costs. In such areas, economic and social integration with the wider community is a major priority. Earlier opportunities for the implementation of multimedia services will be made available by using satellites in regions lacking the adequate infrastructure to exchange all types of information electronically. Satellite networks can provide coverage of large areas rapidly and economically through inexpensive terminals with transmission rates ranging from a few tens of kbps to a few tens of Mbps.

From a user's perspective, standard applications should be supported across all types of networks, and hence the requirement to transport standard network protocols across both space and terrestrial components will be required. Ideally, the user should be unaware of the type of network used to provide the required service or application, be it terrestrial or satellite. The success in recent years of the Internet has resulted in the Transmission Control Protocol/Internet Protocol (TCP/IP) protocol being the most widely used network protocol. Asynchronous Transfer Mode (ATM) is a candidate both as a support for TCP/IP protocol and for dedicated applications above satellite Medium Access Control (MAC) layer. TCP/IP has been designed for terrestrial networks, which are characterised by low delays and error rates. This is unfortunately not true for satellite networks with geostationary satellites, where the one hop (earth station-satellite-earth station) propagation delay is about 250 ms and the bit error rates can become significant in case of fading. However, investigations show that applications like file transfer, database access, remote-login, and e-mail can be well supported on satellite links. The situation is more complex when high interactive or real-time applications are supported (client–server applications, video, voice, etc.). To decrease the delay, satellite networks which use non-geostationary satellites are more attractive.

1.1 Evolution of Satellite Communications

The launch of the first man-made earth satellite SPUTNIK-I by the former Soviet Union in 1957 signified the beginning of the satellite era. This was soon followed by the launch of the first voice communication satellite Explorer-I by the US in 1958. Since then, considerable effort has been spent on the research and development of satellite technologies. In 1963, the first geostationary satellite, Syncom-II, was successfully launched. However, it was the launch of the first commercial geosynchronous satellite that marked the breakthrough in satellite communication for international point-to-point telecommunication applications. In the 1970s, the application of satellite communication extended to direct satellite broadcast and by the 1980s, satellites had been used to provide mobile communications to the maritime sector. By the early 1990s, satellite systems had also been used to provide aeronautical, land-mobile and personal communication services, capable of serving relatively small and inexpensive terminals with transmit and receive capabilities. Such a terminal can either be a mobile terminal with functions similar to that of the GSM or a satellite home dish which is able to receive TV programmes and also to transmit with a satellite return channel, allowing broadband interactive services to be provided via satellite, such as fast internet access.

The need to reduce the terminal size and cost to allow for a mass market in personal communication services imposes additional complexity on board the satellite system and in particular, the requirements in high power levels from the satellite constellation. The complexity of the space segment is directly affected by the choice of orbital types. Specifically, three types of orbits are normally considered: Geostationary Orbit (GEO), Medium Earth Orbit (MEO) and Low Earth Orbit (LEO). Their effects on the complexity on the satellite segment are discussed later.

Following this evolution path, two broad classes of services have been envisaged to be provided by satellite systems:

- *Satellite Personal Communications Services (S-PCS)*. S-PCS systems were first proposed in 1990 by the US-based Iridium consortium led by Motorola and aimed to offer primarily voice services to small, inexpensive pocket-size terminals from a global constellation of LEO satellites. This paved the way for several other S-PCS developments using non-geostationary satellites. Unfortunately, the significant market penetration derived from S-PCS that was anticipated at the start of the decade has failed to materialise since its first introduction in 1998.

- *Broadband Internet type multimedia services.* For advanced broadband multimedia services, communication satellite systems with very high capacity have been proposed, allowing high rate direct user access with low cost interactive user terminals. It is envisaged that such systems will play an important role in the Global Information Infrastructures, especially in the field of Internet access. European proposals for broadband satellite systems include Alenia Spazio EuroSkyWays, Alcatel Skybridge, Matra Marconi WEST, Aerospatiale MEDSAT, SES ASTRA complement.

1.2 EU Initiatives in Satellite Communications

In Europe, the responsibilities of satellite communications R&D are shared among the EU, the National Space Agencies and the European Space Agency (ESA). While the National Space Agencies and the ESA concentrate on costly space technology developments, the EU programmes focus more on satellite-related activities – primarily on equipment and applications.

Most of the EU satellite programmes are pursued with an aim to maximise commonality and interoperability between the satellite and terrestrial network components [1]. This is based on the fact that satellite communications remain a niche market and is expected to play a complementary role to its terrestrial counterpart. As a matter of fact, successful satellite communication systems have illustrated the validity of this approach. For instance, the DVB/MPEG standard adopted in Digital TV broadcasting systems such as DirecTV in the US, Eutelsat and Astra in Europe is a direct result of the picture coding R&D activities pursued in terrestrial networks (although MPEG was not originally developed for sole satellite environment). The mobile satellite industries have also followed such an approach. For example, the Iridium have followed the concepts developed for GSM, and broadband multimedia satellite systems, such as Teledesic or Skybridge, have also considered compatibility with the standards developed for terrestrial networks. Thus, co-ordination between satellite and terrestrial communication industries is of paramount importance to ensure that the standards designed and developed for the terrestrial networks will also be supported in satellite operational environments.

Within the EU COST Action programme, the COST 226 Action has investigated the feasibility of terrestrial network interconnection (such as Local Area Network, LAN, for instance) by GEO systems using transparent satellites and fixed low cost Very Small Aperture Terminals (VSAT) earth stations [2].

Under the RACE II programme, the RACE CATALYST project demonstrated the compatibility of satellite technology with ATM and the terrestrial B-ISDN for LAN interconnection. The equipment developed during the CATALYST project has been able to interconnect ATM test-beds as well as existing networks such as DQDB, FDDI and Ethernet networks, all using ATM. A detailed explanation of the system design and performance is provided in [3–5].

Under the ACTS programme, a total of twelve SATCOM projects were funded, of which the ISIS [6], DIGISAT [7], and the VANTAGE [8] project are of direct relevance to the COST 253 Action. ISIS demonstrated the technical and economical feasibility of interactive services via satellite, in the framework of multimedia applications, utilising Ku band for the forward service and Ka-band for the return interactive link for the support of asymmetric traffic. Similar to the ISIS project, the DIGISAT project developed the system specifications as well as prototypes in order to demonstrate the technical and commercial feasibility of the interactive channel for satellite master antenna television (SMATV). VANTAGE was the follow-on project of CATALYST. VANTAGE united the service flexibility of ATM and access flexibility of satellites to provide a pan-European interconnection service. The VANTAGE platform was utilised by another ACTS project THESEUS [9] to provide inter-connection between remote users and the stock exchanges.

Under the IST programme, several projects are of relevance to the COST253 activities. BRAHMS [10] aims at defining a universal user-access interface for broadband satellite multimedia services for different satellite system implementations, including GEO and LEO constellations. The MULTIKARA [11] project aims at designing and testing innovative multibeam-receiving antenna at around 30 GHz with associated microwave circuits and evaluating the feasibility of such receiver for future in-flight use. GEOCAST [12], although oriented towards the use of geostationary satellites, intends to define the next generation multicast systems by combining multicast services with next generation satellite systems through a progressive and well-adapted strategy. The SATIN7 [13] project aims at developing technologies and architectures for satellite interactive multimedia IP network and services.

Other satellite-related projects have also been funded from the Telematics programme, mainly for the demonstration of services of public interest via advanced satellite communication systems. In particular, the MERMAID project for healthcare applications and the SHARED project, demonstrating advanced video conferencing systems for remote medical assistance in Sarajevo.

1.3 Operating Frequency

The frequency band initially assigned to commercial satellite communications operation is in the C-band (6 GHz uplink, 4 GHz downlink). For mobile satellite services, the ITU has initially allocated the spectrum in the L-band. As the range of systems and services on offer have increased, the demand for bandwidth has resulted in a greater range of operating frequencies, from Ku (e.g. Skybridge) up to Ka band (e.g. EuroSkyway, WEST). Recently, Q/ V band systems have also been planned such as the Motorola M-Star, the Lockheed Martin Q/ V-System and PanAmSat Vstream.

1.4 Technical Considerations

The initial design of communication satellite systems operating at C- and Ku-band deployed a limited number of antenna beams. With the need to operate with small and inexpensive user terminals at high data rates and the requirements to achieve link margins to overcome rain fading, advanced satellite systems entail the use of a large number of narrow spot beams. The problem of rain fading is the most severe for Ka-band operations – most systems are designed to achieved only 99.5% availability. This will limit the range and the quality of services that can be provided for users. The problem is even more serious for systems operating at the Q/V band. The following outlines some technical challenges which need to be considered in satellite communication operations.

1.4.1 Terminal

In order to ensure mass market penetration for broadband multimedia satellite services, user terminals have to be light weight and low cost. Ideally, a terminal with an antenna size in the order of 1 meter and with 10 W of power is required [14]. With systems operating at Ka and V-band, it is difficult to generate radio frequency (RF) power with existing solid-state devices and will be too expensive for silicon fabrication of the up- and down- converters.

Another way to reduce terminal costs is to adopt agreed upon standards. With the need for on-board processing, error-correcting coding schemes and specific protocols to handle services such as ATM over satellite as well as different bit rates for different market segments, it is highly unlikely that common standards can be agreed upon.

1.4.2 Propagation

Attenuation of signal due to rain fading for systems operating at Ka-band is significant. The combined effect of gaseous absorption and rain attenuation will become even worse with systems operating at Q/V-band. In tropical regions, even a 99% availability requires in excess of a 30 dB rain fade margin. Several strategies can be adopted to improve link margin such as the use of increased coding gain, exploitation of uplink (and possibly downlink) power control, operating at reduced bit rates and at lower operating frequency band or the use of diversity techniques.

The two most common approaches are to improve coding gain and to exploit uplink power control [14]. Improved coding gain can be achieved by the use of concatenated coding and this is regarded as a necessity for Q/V-band operation. Uplink power control increases the uplink earth station power during rain. This requires the monitoring of received signal strength. For systems with on-board processing capabilities, a simple switch to high power may be sufficient. However, the use of power control may require an increase of 10 dB or more in power; this may in turn require switching from a solid-state amplifier to a travelling wave tube which may become costly.

1.4.3 Orbital Types

Up until in the 1990s, geostationary satellites had been used as the sole basis for the provision of satellite real-time communication services. The major advantage of GEO satellites are their unchanging position with respect to the earth surface, thus no control overhead is required to track the satellites. They require the least user co-operation, allowing services to be globally and continuously provided with only three satellites in orbit. The drawbacks are the high cost in launching the satellites into the geostationary orbit, the long inter-satellite link distance, the high on-board power required, and the long propagation delays. Since geostationary satellites remain to be stationary with respect to the earth, a lot of satellite capacity will be wasted in covering areas where no or little traffic are generated. It is desirable to allow feasibility in assigning different amounts of capacity to each beam or to employ scanning beams whose dwell time can be adjusted dynamically according to the amount of traffic in the coverage area.

Another extreme is to employ LEO satellites. LEO satellites altitudes vary between 750–2000 km and have received a lot of attention during the 1990s for their low latency and their less stringent high power requirement than that of GEO satellites. Their low latency has overcome the limitations introduced by the TCP/IP protocol on throughout for GEO satellites. However, LEO satellites provide time-dependent coverage over a particular area. As a result, multisatellite constellations of more than 30 satellites are required for global, continuous coverage. Phased array antennas are usually deployed such that satellite beams can remain fixed on their service areas before the satellite hops to a new service area. This would allow a user to remain in the same beam for as long as possible. The dynamic nature of the satellite orbits require handover between satellites during a call. A large number of gateways is also required to support a global network if inter-satellite links are not employed. Furthermore, satellites need to be tracked continuously.

MEO satellites offer lower tracking rate than that of LEO satellites. Some systems employ beam steering or switching to extend the user dwell time in a given beam while others allow beams to be fixed relative to the satellite and sweep over the ground. Since the altitude of a MEO satellite lies typically between 10,000–20,000 km, multisatellite constellations of 10–20 satellites are required for global continuous coverage. MEO systems typically provide users with higher look angles, reducing the path loss and attenuation through the atmospheric gases and rain path. Although handover between satellites during a call is still required, the frequency is significantly reduced in comparison with LEO satellites. The larger coverage area also allows a less complicated ground network to be supported.

1.4.4 On-Board Routing

Most of the global multimedia and data distribution systems employ some form of on-board routing in order to route traffic from beam to beam, and in many cases from satellite to satellite using Inter-Satellite Links (ISLs). On-board routing can be implemented in one of the following ways [14]:

1. The traffic in each beam can be sub-divided into several channels. A static switch matrix in the IF stages of the transponder to route traffic streams to various beams.
2. Connect each uplink beam cyclically to a large number of downlink beams for a brief period by using Satellite Switch Time Division Multiple Access (SSTDMA).
3. Arriving carriers are demultiplexed, demodulated and decoded into raw packets, which are then routed to the appropriate port using a baseband digital switch.

Although the first approach is the simplest to implement, the static nature of the switch matrix does not allow flexibility, resulting in congestion in some switch paths whilst others may remain under-utilised for most of the time. The second approach limits the choice of multiple access techniques and may increase the cost of the earth station. The third approach is suitable if the number of beams is high. The drawbacks are the high risk and the increase in the satellite weight and cost because a large on-board processor is required, which can consume a lot of satellite power.

1.4.5 Traffic Management

Internet applications tend to generate asymmetric traffic such that a moderate data rate is required in the uplink whilst data is transmitted at a much higher speed. Furthermore, most of the internet traffic will be concentrated in developed countries such as EU countries, US and

Japan. Satellite capacity should be managed more effectively to take into account the traffic asymmetry and the variation in traffic density around the globe. The adoption of ISLs to route traffic to designated areas is seen to be advantageous.

1.5 Objectives and Activities of COST253

1.5.1 Objectives

The work under COST Action 253, relying on concurrent efforts under other international projects, has focussed on assessing the applicability of early proposed mechanisms within satellite systems interworking with terrestrial communication networks. It has also focussed on analysing specific internal mechanisms able to cope with or to benefit from the characteristics of satellite systems, within the emerging and evolving standardisation framework.

The underlying objective of COST Action 253 is to make satellites active and dynamic components of global networks incorporating both satellite links and terrestrial networks. In particular, non-geostationary satellites have been considered with a view to minimise transmission delay restrictions, aiming at providing an efficient communications system particularly for areas with inadequate ground infrastructure.

In this context, the goals of the action are as follows:

- Identify problems related to satellite motion and impacts on satellite gateway and transmission system design.
- Explore whether present protocols or those elaborated within the EU frameworks for GEO satellite networks are suitable or need to be modified and the mechanisms for such modifications.
- Develop software for system simulation, dimensioning and performance evaluation for systems for network interconnection through non-geostationary satellites. The implementation of hardware systems and software is foreseen to provide a test-bed for verifying the transmission and access schemes as well as the performance of the protocols and applications.
- Define a limited set of satellite system generic reference architectures and configurations, coping with private or access networking scenarios, and including constellation of multi non-geostationary satellites.
- Assess recommended schemes and mechanisms in the area of traffic, quality of service (QoS) and resource management and their applicability within the defined satellite systems.
- Analyse and recommend specific functions to cope with the characteristics of the considered satellite systems.
- Support the implementation of the main defined schemes within a simulated, and if possible, a satellite system.
- Ensure the distribution of results through papers, workshops, etc.

1.5.2 Action Activities

To achieve the objectives and goals outlined above, two main work areas have originally been identified: *Network integration by non-geostationary satellites* and *Traffic, quality of service and resource management techniques.* Within these two work areas, several aspects have been included such as the transmission schemes, access methods for multi-satellite environment, fade-countermeasures and transmission frequency considerations, internetworking, gateways and traffic modelling.

According to the two main work areas, the activities of COST253 has been allocated to six different working groups:

- WG1: Traffic characterisation
- WG2: Definition of network organisation
- WG3: Transmission schemes
- WG4: Networking
- WG5: Security issues
- WG6: Performance evaluation and recommendations

1.5.2.1 WG1: Traffic Characterisation

Under this workgroup, different types of services for different network architectures and their source modelling were studied. Furthermore, traffic in/out of the satellite node for passive or regenerative satellites (switching) for various channel access techniques were investigated together with an end-to-end traffic model being justified by simulation work.

1.5.2.2 WG2: Definition of Network Organisation

The objective of this workgroup is to determine the number, size and type of terminals to be interconnected in a reference constellation. The effects of ISL/On-Board Processing (OBP)/ On-Board Switching (OBS) on the network performance were also studied. The output of WG2 was used to define a more detailed network and subsystem architecture.

1.5.2.3 WG3 Transmission Schemes

Due to the significant relative movement between the ground station and the satellite, Döppler effects resulting in frequency shifts on the carrier frequency and clock slips on the demodulated digital signal cannot be neglected. The transmission scheme has to be robust in order to cope with these frequency shifts. The tracking range of the demodulator loops have to be wide enough for synchronisation. In order to cope with cycle slips, sufficient buffering has to be provided. On the other hand, the buffering must be optimised such that no significant delay be contributed. Furthermore, advanced protocols have to be provided specifically for such a satellite environment.

For satellite systems operating at Ku-band or above, the fading effects can be significant. In order to provide reliable communications without large fade margins, efficient fade counter-measures have to be applied. Various transmission and coding techniques have been studied in this work group in order to overcome the effect of fading.

1.5.2.4 WG4 Networking

In this workgroup, the problems of internetworking between heterogeneous networks, the

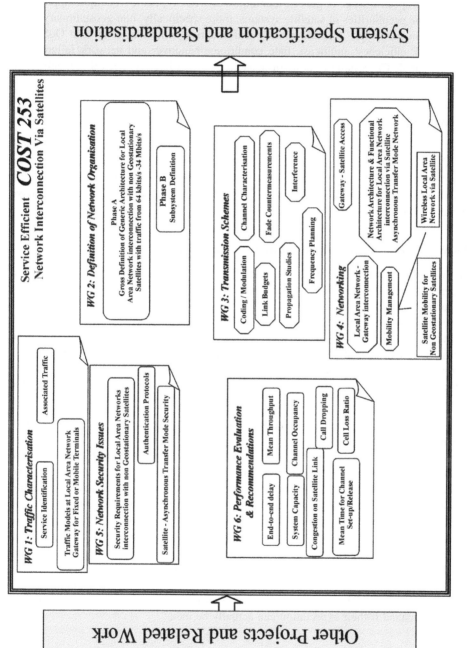

Figure 1.1 Interconnections between different working groups

application of advanced techniques on bridging and routing have been studied. Taking into account the results obtained in WG1 and WG2, this workgroup focused on analysing and recommending specific usage of generic techniques as well as specific techniques appropriate to deal with the peculiarities of satellite systems, more specifically, non-geostationary satellite systems. In particular, dynamic routing and adaptive control to cope with QoS, load balancing, link efficiency, multicast, fade conditions, handover and network failure have been investigated. The reliability issue of a satellite system has also been thoroughly dealt with.

1.5.2.5 WG5 Security Issues

This workgroup studied issues on threats and security requirements. Suitable encryption and digital signatures systems have also been identified. Authentication and encryption key exchange protocols were investigated in relation to satellite-ATM security implementation.

Other issues such as Trusted Third Parties, network management security and billing have also been studied.

1.5.2.6 WG6: Performance evaluation and Recommendations

The activity of this workgroup was mainly concerned with performance measurement of a heterogeneous network where multimedia connections are treated as a bundling of single-media connections.

Figure 1.1 shows the interconnection amongst the working group.

In addition to these workgroup activities, several short-term scientific missions have been supported. Through these short term missions, an open platform network simulator, GALILEO, for the evaluation of satellite system performance has been developed. In order to optimise the simulation time, the following approaches have been taken in developing the software:

- Near real systems were simulated and the performance of such systems was evaluated with a realistic set of parameters.
- Blocks of simulation software have been set up to serve as building blocks for more elaborate simulators as experience was gathered, aiming at designing, developing and validating a modular simulation and emulation environment. This network simulator would allow performance evaluation of future satellite networks by providing a valuable insight into the achievable performance of future satellite systems, in terms of quality of service, mainly from a network perspective standpoint. By using the simulation/emulation environment, system designers will be able to verify how different design solutions would impact on the system performance so minimising the risks from the early stage of the design phases. On the other hand, potential satellite service providers and investors could perform comparative analyses among different candidate system solutions in order to better understand if their expectation can actually be met.

1.6 Outline of Contents

Although not all the topics listed in the six working group activities were fully covered, a large amount of work has been done and considerable number of papers covering specific items have been published in relation to this COST action (see Appendix B). Following this chapter, this book is organised into six other chapters covering the results of the work performed in this Action. Some results from different working groups were aggregated together. In the following, the mapping between working groups and respective chapters are presented.

1.6.1 Chapter 2: Appropriate Traffic Generators for the Simulation of Services Supported by Non-GEO Constellations

WG1 studied the characterisation/modelling of aggregated traffic at the earth station or at the on-board switch of the LEO constellation. An overview of source modelling and service characterisation was provided as a basic platform, but the research mainly focused on self-similar modelling. Research on geographic traffic modelling was carried out to cope with the characteristic of non-uniform loading of the LEO constellations.

The work done by the WG1 team is reflected in Chapter 2, where service categories and requirements are classified, a source activity is characterised (Markovian and Regression models), and models for aggregated traffic are presented.

1.6.2 Chapter 3: Transmission Schemes

The radio interface design in the broadband satellite systems is a major factor determining the efficiency of the systems in the utilisation of the satellite system resources such as bandwidth and power. Utilisation of advanced techniques in the radio interface can bring important system enhancements. Within WG3 some existing schemes and promising options were analysed.

The results achieved by the WG3 team are reflected in Chapter 3. They are limited to an analysis of modulation schemes, coding schemes and Döppler frequency shift correction. For modulation techniques and coding techniques, a general survey of possible schemes have been presented, and then a study of one particular scheme in more detail has been carried out. For the Döppler shift correction, three techniques have been studied in detail for comparison; all three techniques are based on correction in the receiver. For all the particular studies, the results are based on Monte Carlo simulations. The interference problems also were subject to evaluation, as the strict international regulations must be respected by all systems. The crowdedness of the Ku-band imposes interference control. The work presented in Chapter 3 is related to the modellisation of interference created by user signals at the satellite, assuming a CDMA-based system.

1.6.3 Chapter 4: Networking

This chapter assembles the major results of the WG2, WG4 and WG5 and WG6 teams. The main objective of WG2 was to define a general network architecture to interconnect terrestrial networks with traffic ranging from a few tens of kbps to several hundred Mbps, using non-geostationary satellites. The satellite network entity options, which have to be taken into account in the design of a satellite constellation, were analysed. Then the key satellite payload architecture options were identified, and their advantages/disadvantages discussed. In order to be able to choose an optimal satellite constellation and satellite architecture to efficiently interconnect terrestrial networks, recent non-GEO broadband proposals were compared. The selected satellite network entity options, taken by the COST253 Managment Committee, use ISLs and OBS. ISLs were selected because they provide some independence from the terrestrial infrastructure allowing end-to-end routing over the space segment. ISLs require OBS, which was selected because of the flexibility it provides in routing and multicast.

Work undertaken in WG4 mainly addressed routing strategies and traffic congestion control mechanisms. Concerning the routing strategies, a simulation tool was developed, having the goal to study the effects of the different routing algorithms on the performance of the network (such as the blocking probability during connection set-up, the load distribution in the network, the properties of the routes, and the complexity of the routing algorithm), and to study the effects of Link State Packet data unit (LSPdu), broadcast policies on the network load, and the blocking probability during set up.

In Chapter 4 several network control issues are considered: the use of non-geostationary satellites to provide terrestrial network interconnection, the network resource control aspects, the routing strategies, call control functions and multicasting techniques. Results from different studies are presented. A detailed description on the reliability requirements and techniques for satellite system is also presented. Finally, the security requirements and their implementation issues are discussed.

WG5 had to face with problems related to the user-satellite network authentication, information privacy and satellite-ATM security. Data encryption and digital signature techniques were studied and elaborated, and their implementation in both hardware and software examined.

1.6.4 Chapter 5: Evaluation Tools

WG6 had to collect, in a certain sense, the results of the previous WGs, and measure the performance of a heterogeneous network where multimedia connections are treated as a bundling of single-media connections, of which satellites are one possibility. Simulation proved to be the only possibility to carry out this work, organising it in groups of parameters which had to be tested according to both the characteristics of the satellite network (such as different satellite allocation policies, different number of satellites per orbit, different types of faults in the satellites of the constellation, etc.), and other choices relevant to the terrestrial network.

As different aspects of the non-geostationary satellite network had to be studied, different ad-hoc simulation tools have been used, and a new one was studied whose aim is (because its development is still in course) to provide a test-bed for studying various aspects of communications using satellite constellations.

Chapter 5 is devoted to the presentation of different evaluation tools.

References

[1] B. Barani, 'Satellite Communications in the European Union R&D Programmes: an Overview' *Proceedings of IEE Colloquium on EU's Initiatives in Satellite Communications*, 1/1–1/8, May 1997, London.
[2] 'COST 226 – Integrated Space/Terrestrial Networks', Final Symposium, Budapest, Hungary, 10–12 May, 1995.

[3] P. Polese, R. Mort and L. Combarel, 'Satellites in UMTS and BISDN: status of activities and perspectives' *Electronics and Communication Engineering Journal*, 297–303, 1994.

[4] M.H. Hadjitheodosiou, P. Komisarczuk, F. Coakley, B.G. Evans and C. Smythe, 'Broadband island interconnection via satellite performance analysis for the RACE IICATALYST Project', *International Journal of Satellite Communications*, Vol. 12, pp. 223–238, 1994.

[5] Z. Sun, T. Ors and B.G. Evans, 'Interconnection of broadband islands via satellite-experiments on the RACE II CATALYS Project', *IEEE Globecom Workshop on Transport Protocols for High-Speed Broadband Networks*, 1996.

[6] C. Dorides and F. Carducci, 'ISIS project – An Open Platform for Multimedia Interactive Services' *Proceedings of IEE Colloquium on EU's Initiatives in Satellite Communications*, 6/1–6/12, May 1997, London.

[7] J. Sesena, H. Prieto, A. Molina and M.J. Sedes, 'The Satellite Role in the Interactive Broadcasting Era – Application to SMATV (DIGISAT Project)' *Proceedings of IEE Colloquium on EU's Initiatives in Satellite Communications*, 5/1–5/10, May 1997, London.

[8] S. Chellingsworth, 'VANTAGE–VSAT ATM Network Trials for Applications Groups Across Europe' *Proceedings of IEE Colloquium on EU's Initiatives in Satellite Communications*, 8/1–8/5, May 1997, London.

[9] F. Dachet, M.N. Sauvayre and F. Capobianco, 'Remote Trading by Satellite-ATM: Implementation Issues in the Scope of the Theseus Project' *Proceedings of IEE Colloquium on EU's Initiatives in Satellite Communications*, 3/1–3/6, May 1997, London.

[10] http://www.cselt.it/brahms

[11] http://www.arttic.com/projects/multikara

[12] http://www.cordis.lu/ist/projects/99-11754.htm

[13] http://www.cordis.lu/ist/

[14] J.V. Evans, 'The US filings for multimedia satellites: a review' *International Journal of Satellite Communications*, Vol. 18, No. 3, pp. 121–160, 2000.

2

Appropriate Traffic Generators for the Simulation of Services Supported by Non-GEO Constellation

Provision of the requested Quality of Service (QoS) and efficient utilisation of resources are some of the objectives of COST253. In order to achieve these challenging tasks it is necessary to analyse and/or simulate such a network before actually implementing it. Modelling services using appropriate source and traffic models are therefore very important to enable simulations and analysis to be performed with the selected network architecture.

Source modelling is used to mimic the behaviour of a source. Traffic modelling on the other hand focuses on aggregated traffic patterns. Multiplexed models will capture the effects of statistically multiplexing bursty sources and will predict to what extent the combination of bursty stream is smoothed. Hence traffic models will be used for designing connection admission control algorithms and traffic engineering the network. The most important application of source models is in predicting the QoS that a particular application might experience under various network conditions. In order to assess the QoS provided to individual services, source models and aggregated source models (traffic models) will be used extensively.

A number of traffic models are already known from the traditional fixed and mobile network planning and they can be classified in:

1. Short-range dependent models, which have a correlation structure that is significant for relatively small lags; and
2. Long-range dependent processes, which have a significant correlation even for large lags.

Traffic models can be based on analysis, simulation and measurements. Measurements are used to verify the accuracy of the simulation model, which is usually based on analysis.

Source characterisation at the macro level is defining the source traffic characteristics and its QoS requirements. The traffic characteristics of an application are the minimum set of parameters that a user can be expected to declare whilst providing the network with as much information as possible to effectively control network traffic and achieve high resource utilisation.

After defining various parameters which can be negotiated between user and network, service categories supported by ATM networks are identified. Each service category can support various services depending on which traffic parameters can be declared and which QoS guarantees are required by the user. The statistical behaviour of generic traffic services such as voice, video, data and multimedia are provided. Then the influences on traffic characteristics are discussed and the criteria in selecting source models for traffic sources are explained. Finally a review of widely used source models and traffic models is provided.

2.1 Source Traffic Parameters and Descriptors

Source traffic parameters are used to describe traffic characteristics of a source. They may be quantitative or qualitative (e.g. telephone, videophone). For an ATM connection, traffic parameters are grouped into a *source traffic descriptor*, which in turn is a component of a *connection traffic descriptor*.

A *source traffic descriptor* is the set of traffic parameters of the ATM source. It is used during the connection set-up to capture the intrinsic traffic characteristics of the connection requested by a particular source. The set of traffic parameters in a source traffic descriptor can vary from connection to connection. A *connection traffic descriptor* characterises a connection at the User Network Interface (UNI). It consists of:

- Source traffic descriptor
- Cell Delay Variation Tolerance (CDVT)
- Conformance definition

The connection traffic descriptor is used by the network during connection set-up to allocate network resources and derive parameters for Usage Parameter Control (UPC). The conformance definition is used by the UPC to distinguish conforming and non-conforming cells without ambiguity.

An important issue is the set of traffic parameters to include in the source traffic descriptor. All parameters should be simple to be determinable by the user, interpretable for billing, useful to Connection Admission Control (CAC) for resource allocation, and enforceable by UPC. The set should be small but sufficient for the diverse types of traffic in B-ISDN. Some proposed source traffic parameters, which will be explained in detail, are:

- Peak Cell Rate (PCR) and Cell Delay Variation Tolerance (CDVT)
- Sustainable Cell Rate (SCR) and Maximum Burst Size (MBS)
- Mean burst duration (T_{on})
- Intrinsic Burst Tolerance (IBT)

2.1.1 Peak Cell Rate and Cell Delay Variation Tolerance

The Peak Cell Rate (PCR) of the ATM connection is the inverse of the minimum interarrival time T between two cells on a transmission link. It specifies an upper bound on the traffic that can be submitted on an ATM connection [17]. The ATM Forum and International Telecommunications Union-T define the PCR and CDVT using the *Generic Cell Rate Algorithm* (GCRA) and *equivalent terminal* model [17,18]. The reason for variation in the cell delay is that ATM layer functions (e.g. cell multiplexing) may alter the traffic characteristics of ATM connections by introducing CDV. When cells from two or more ATM connections are multiplexed, cells of a given ATM connection may be delayed while cells of another ATM connection are being inserted at the output of the multiplexer. Similarly, some cells may be delayed while physical layer overhead or Operation and Maintenance (OAM) cells are inserted. Consequently with reference to the *peak emission interval, T*, (i.e. the inverse of the contracted peak rate), some randomness may affect the inter-arrival time between consecutive cells of a connection. The upper bound on the 'clumping' measure is the CDVT. The CDVT at the public UNI, is defined in relation to the PCR according to the GCRA (T, τ_{UNI}).

For the time being two extreme cases of characterising the CDVT [17] have been identified:

1. *Loose requirements on CDVTA*: large amount of CDV can be tolerated. In this case, only the specification of the maximum value of CDVT τ_{MAX} that can be allocated to a connection is envisaged. τ_{MAX} is intended as the maximum amount of CDV that can be tolerated by the user data cell stream.
2. *Stringent requirement on CDVTA*: connection should not be denied because of the required CDVT, if this CDVT requirement is less than or equal to τ_{PCR}, given by:

$$\frac{\tau_{PCR}}{\Delta} = \max\left[\frac{T_{PCR}}{\Delta}, \alpha\left(1 - \frac{\Delta}{T_{PCR}}\right)\right] \tag{2.1}$$

where T_{PCR} is the peak emission interval of the connection (in seconds), Δ is the cell transmission time (in seconds) at the interface link speed, α is a dimensionless coefficient (suggested value is 80 [17]).

2.1.2 Sustainable Cell Rate and Intrinsic Burst Tolerance

SCR is an upper bound on the *average rate* of the conforming cells of an ATM connection, over time scales which are long relative to those for which the PCR is defined. The Intrinsic Burst Tolerance (IBT) [17] specifies the maximum burst size at the PCR or in other words the maximum deviation from the *average rate*. These parameters are intended to describe Variable Bit Rate (VBR) sources and allow for statistical multiplexing of traffic flows from such sources.

The SCR and IBT traffic parameters are optional traffic parameters a user may choose to declare jointly, if the user can upper bound the average cell rate of the ATM connection. To be useful to the network provider and the customer, the value of the SCR must be less than the PCR. The SCR and the IBT (denoted as τ_{IBT}) are defined by the GCRA (T_{SCR}, τ_{IBT}). SCR and IBT belong to the ATM traffic descriptor [17]. Translation from the Maximum Burst Size (MBS) to τ_{IBT} will use the following rule:

$$\tau_{IBT} = \lceil(MBS - 1)(T_{SCR} - T_{PCR})\rceil \text{ seconds} \tag{2.2}$$

where [x] stands for the first value above x out of the generic list of values.

If the user has the knowledge of τ_{IBT} rather than of the maximum burst size, than the following rule applies:

$$MBS = 1 + \lfloor[\tau_{IBT}/(T_{SCR} - T_{PCR})]\rfloor \text{ cells} \tag{2.3}$$

where [x] stands for rounding down to the nearest integer value.

2.1.3 Mean Burst Duration

The mean burst period (T_{on}) is defined as the average time the source is transmitting cells at the peak rate. This parameter is widely used for bursty sources.

2.1.4 Burstiness

Burstiness (β), following the ITU-T definitions, corresponds to the ratio of the peak-to-average traffic generation rate (β = PCR/SCR).

2.2 Quality of Service Parameters

QoS is measured by a set of parameters characterising the performance of an ATM layer connection. These QoS parameters (referred to as network performance parameters by ITU-T) quantify end-to-end network performance at the ATM layer.

Six QoS parameters are identified by the ITU-T and ATM-Forum which correspond to a network performance objective. Three of these may be negotiated between the end-systems and the networks. One or more values of the QoS parameters may be offered on a per connection basis, corresponding to the number of related performance objectives supported by the network. Support of different performance objectives can be done by routing the connection to meet different objectives, or by implementation-specific mechanisms within individual network elements. The following QoS parameters are negotiated:

- *Cell Loss Ratio (CLR)*: CLR is the ratio of total lost cells to total transmitted cells in a connection.
- *Cell Transfer Delay (CTD)*: This is defined as the elapsed time between a cell exit event at the measurement point 1 and the corresponding cell entry event at measurement point 2 for a particular connection.
- CDV: Two performance parameters associated with CDV are defined. The first parameter, One-Point CDV, is defined based on the observation of a sequence of consecutive cell arrivals at a single Measurement Point (MP). The second parameter, Two-Point CDV, is defined based on the observation of corresponding cell arrivals at two MPs that delimit a virtual connection portion. Ref. [18] provides more details on CDV measurements.

The following QoS parameters are not negotiated:

- *Cell Error Ratio (CER)*: CER is the ratio of total errored cells to the total of successfully transferred cells in a connection.
- *Severely Errored Cell Block Ratio (SECBR)*: SECBR is the ratio of total severely errored cell blocks to total cell blocks in a connection.
- *Cell Misinsertion Rate (CMR)*: CMR is the total number of misinserted cells observed during a specified time interval divided by the time interval duration.

Further information on ATM layer QoS may be found in ITUT Recommendation I.356.

2.3 ATM Service Categories

Services provided at the ATM layer, consists of different service categories which will be explained in this section.

2.3.1 Constant Bit Rate (CBR) Service Category

The CBR service category is used by connections that request a static amount of bandwidth that is continuously available during the connection. This amount of bandwidth is characterised by the PCR. CBR service is intended to support real-time (rt) applications requiring tightly constraint delay variation but is not restricted to these applications. Typical examples of CBR services include voice, video, and audio.

In the classical Synchronous Transfer Mode (STM) networks, the fluctuating information rate must be converted into a CBR, namely the rate at which this STM network is operating. For instance, 64 kbit/s or 2 Mbit/s in N-ISDN.

CBR traffic is easy to manage in the network, since constant bandwidth is reserved for each CBR connection throughout its duration, independent of whether the source is actively transmitting or in a silent state. This is, however, an inefficient use of the transmission bandwidth. In particular, since the amount of information generated by most applications varies over time it is possible to reserve less bandwidth in the network than the application's peak bit rate, thereby allowing more connections to be multiplexed and increasing the resource utilisation. In initial deployments, a large portion of traffic in ATM networks is expected to be CBR voice, video and audio. As time evolves, designers will have a better understanding of the dynamics of VBR traffic and be able to design efficient techniques to manage VBR traffic in the network, thereby achieving high resource utilisation.

2.3.2 Variable Bit Rate (VBR) Service Category

The traffic generated by a typical source, in general, either alternates between the active and silent periods and/or has a varying bit rate generated continuously. Furthermore, the peak-to-average bit rate (burstiness) of a VBR source is often much greater than one. Presenting VBR traffic to the network as CBR traffic means buffering, or rather artificially controlling its bit generation rate, which has the drawback of underutilisation of network resources and QoS degradation. Although doing so simplifies the network management task, it is more natural to provide VBR service to VBR sources and thereby provide a better service and a framework to achieve higher resource utilisation. ATM networks offer this opportunity, thus the limitations of working at CBR disappears. VBR connections are characterised in terms of PCR, SCR and IBT.

The VBR service category is usually divided into two categories namely: *real-time Variable Bit Rate (rt-VBR)* service category and *non-real-time Variable Bit Rate (nrt-VBR)* Service Category. The rt-VBR service category is intended for rt-applications which require tight constrained delay and delay variation. The nrt-VBR Service Category on the other hand is intended for nrt-applications and no delay bounds are associated with this service category.

2.3.3 Available Bit Rate (ABR) Service Category

Many applications, mainly handling data transfer, have the ability to reduce their sending rate if the network requires them to do so. Likewise, they may wish to increase their sending rate if there is extra bandwidth available within the network. This kind of applications, which do not require bounds on delay and delay variation are supported by the ABR service category.

A rate-based flow control was specified [17] which supports several types of feedback to control source rate in response to changing ATM layer transfer characteristics. This feedback is conveyed to the source through specific control cells called Resource Management (RM) cells. It is expected that an end-system that adapts its traffic in accordance with the feedback will experience a low cell loss ratio and obtains a fair share of the available bandwidth according to a network specific allocation policy.

On the establishment of an ABR connection, the end system specifies both a maximum and minimum required bandwidth. These are called Peak Cell Rate (PCR) and Minimum Cell Rate (MCR), respectively. The bandwidth available from the network may vary, but is guaranteed not to become less than the MCR.

2.3.4 Unspecified Bit Rate (UBR) Service Category

The UBR service category is intended for nrt-applications like traditional computer communication applications, such as file transfer and e-mail.

UBR service does not specify traffic related service guarantees. No numerical commitments are made with respect to the CLR experienced by a UBR connection, or as to the Cell Transfer Delay (CTD) experienced by cells on the connection. Congestion control for UBR may be performed at a higher layer on an end-to-end basis. The UBR service is indicated by use of the *best effort indicator* in the ATM *user cell rate information element*. Even if the PCR is not enforced it is still recommended to have the PCR negotiated, so that the source can discover the bandwidth limitation of the connection.

Recently there have been proposals to guarantee a minimum bandwidth to UBR sources. This service class has been proposed to be able to support applications like TCP/IP which are loss sensitive and has been called UBR+ or Guaranteed Frame Rate (GFR) [30]. The UBR+ service class is similar to ABR without feedback control.

This Section has defined the ATM service categories. Table 2.1 provides a list of ATM QoS and traffic parameters and identifies whether and how these are supported for each

Table 2.1   List of ATM service traffic categories parameters

Source parameters		CBR	rt-Vt-VBR	UBR	ABR
PCR and CDVT	Specified			Specified	Specified
SCR, MBS, CDVT	N/A	Specified		N/A	N/A
MCR	N/A			N/A	Specified
CDV	Specified	Specified	Unspecified	Unspecified	Unspecified
Maximum CTD	Specified	Specified	Unspecified	Unspecified	Unspecified
CLR	Specified			Unspecified	Specified

Table 2.2   Provisional QoS network performance objectives[a]

	CTD	2-pt. CDV	CLR	CER	CMR	SECBR
Default (objectives)	No default	No default	No default	4×10^{-7}	1/day	10^{-4}
Class 1 (stringent)	400 ms	3 ms	3×10^{-7}	Default	Default	Default
Class 2 (tolerant)	U	U	10^{-5}	Default	Default	Default

[a] 'U' means unbounded. When the objective of a parameter is specified as being 'U' performance with respect to the parameter may, at times, be arbitrarily poor.

service category. Table 2.2 provides the guaranteed network performance objectives of the network for a specific traffic class as recommended by the ITU-T Rec.I.356.

2.3.5 Guaranteed Frame Rate (GFR) Service Category

The Guaranteed Frame Rate (GFR) service is UBR with some level of service guarantees. GFR requires minimal interactions between users and ATM networks, but the simplicity of the service specification for users does come at a cost in terms of requirements imposed on the network in order to efficiently support GFR. However the cost of these requirements is far outweighed by the potential benefits of making ATM technology more attractive to a broad range of users (in particular Internet users).

2.4 Statistical Behaviour of Traffic Sources

2.4.1 Voice

The statistics of a single voice source are composed of two phases and they normally depend on the technique of voice coding that is being used. The two periods are the active period and the silent period.

The POTS (Plain Old Telephony Service) has been using a fixed bandwidth digital channel at 64 kbit/s. Modulation techniques such as adaptive differential pulse code modulation can be used to compress voice information to a constant bit rate with lower bandwidth requirements.

CBR voice in ATM networks is transmitted with AAL type 1 using the pulse code modulation technique. Recommendation G.711 [19] specifies 64 kbit/s CBR voice. When voice signals are coded with a variable bit rate an active period of a voice source corresponds to a talk spurt, whereas a silent period corresponds to speech silence duration. The silent periods constitutes 60–65% of the transmission time of voice calls in each direction. More specifically, the average active and silent periods are measured to be respectively equal to 352 and 650 ms [220. Furthermore, in a normal conversation the active period fits the exponential distribution reasonably well while the duration of the silent periods is less well approximated by the exponential distribution [21]. Nevertheless, the most frequently used models of voice sources in the literature assume that the duration of both active and silent periods are exponentially distributed. A single voice source can be modelled by an Interrupted Poisson Process (IPP) or by the on–off model. Multiplexed voice sources are best modelled by a Markov Modulated Poisson Process (MMPP).

2.4.2 Video

A promising service of ATM networks is video communication. It can be divided into still picture and motion picture video traffic. The investigation of video statistics started in the 1970s, but still little is known about the statistics for the arrival process of cells containing video information coded at high bit rates. Video is quite different than voice or data in that its bit streams exhibit various types of correlation's between consecutive frames. Video images have the following statistical components (which are dependent on the type of codec):

- *Line Correlation*: occurs when data at one part of the image is highly correlated with data on the same part of the next line (spatial correlation).
- *Frame Correlation*: data at one part of the image is highly correlated with data on the same part of the next image (temporal correlation).
- *Scene Correlation*: occurs because sequences of scenes may, to a greater or less extent, be coincidentally correlated with each other.
- *White Noise*: is a memoryless process and is uncorrelated.

Non-frame buffered video codecs have all four of the correlation's, whilst frame buffered video codecs (frames all always buffered before being sent) only have scene and white noise correlation [22]. Scene correlation's can be reduced by multi-frame buffering.

Due to the various correlation's that video traffic exhibits it is inadequate just to measure the burstiness of video traffic. The following list summarises some desirable qualities for new measures:

- The measure should not yield just statistical values, but values that capture the characteristics of the rate variation over time.
- The measures must be capable of evaluating the statistical multiplexing effect.
- The measures should allow easy modelling of video information sources.

The following measures have been proposed [23] to fulfil these kinds of conditions:

- *Bit Rate Distribution*: The distribution and the probability density distribution of the encoded bit rate evaluated in single frame units. Along with the average bit rate and the variance, they are quite adequate for approximating the required capacity.
- *Autocorrelation Function*: The autocorrelation function is a convenient measure for expressing the nature of temporal variations.
- *Coefficient of Variation*: In order to express such phenomena as the signal delays that arise when a signal is buffered, the coefficient of variation is used, as a measure to investigate the multiplexing characteristics when variable-rate signals are statistically multiplexed.
- *Distribution of Scene Duration*: The probability density distribution of intervals between scene changes.

Various models have been proposed to model video sources. In applications with *uniform-activity-level scenes*, the change in the information content of consecutive frames is not significant. A typical application of this form is a video telephone where the screen shows a person talking. In general, correlation's in video services with uniform activity levels last for short duration and decay exponentially with time. A first-order Autoregressive (AR) model is proposed in [24]. Another continuous-state AR model which is found to be quite accurate compared with actual measurements, is proposed in [25]. It has however to be noted that these models are not convenient for queuing analysis, but mostly used in simulation studies. In order to evaluate regions of extremely low probability (like cell loss), Markov models can be used.

The observation that intrascene bit-rate variations are smooth and that their sum should not exhibit sudden jumps was used by [25], to model the video source as a continuous time, discrete state Markov model. This is a type of birth–death Markov model, and only transitions to adjacent states are possible.

In applications with *non-uniform activity-level scenes* like motion video, frames of high-activity scenes and scene changes contain large amounts of data followed by frames that contain less data. In addition to the short-term fast decaying correlation's (temporal correlation's) of uniform activity scenes, there is a long-term slow decaying correlation in the amount of information generated per frame, that occurs at times of scene changes. The Autoregressive Moving Average (ARMA) model is proposed [26] to take into account the two types of correlation that occur in non-uniform activity-level scenes. The ARMA arrival processes are used in Monte Carlo simulations to estimate the probability-distribution function of the queuing delay and the mean and variance of the inter-departure time seen by the arriving cell. However, these models cannot be used in the numerical and analytical analysis of queues.

A two-dimensional, continuous time Markov model which is shown in Figure 2.6 [27], is a generalisation of the model developed in [25] for uniform activity scenes. In two dimensions, it is now possible to model the bit-rate fluctuations in consecutive frames to include jumps to the higher or lower bit rates, thereby modelling the correlation at scene changes.

The MMPP can be used to model the cell arrival process from video sources. The inter-scene transition is given by a Markov chain. This model views bit-rate variations as changes in the number of packet arrivals. Furthermore, for ease of analysis, all distributions (the scene change interval and state persistence-time distribution) are assumed to be exponential distributions.

The MMPP can also be used when N independent video sources are multiplexed. Since the scene change interval distribution of the various sources are assumed to be exponential distributions, the scene change interval distribution of the multiplexed model will be an exponential distribution with an average scene change interval of $1/N$ of that of a single video source.

2.4.3 Data

The term data is used for any application that uses coded text, that is, any application that is not voice, audio, video or still image. Despite the fact that data networks have been operational for a number of decades, traffic characteristics of some data sources are not well understood.

The main difficulty arises due to the fact that there is no typical data connection. Large amounts of data are transmitted in a file transfer on a rather continuous basis during the duration of the connection, whereas only a few hundred bytes are generated by an e-mail. Furthermore, data connections are not generally established between two users, but between groups of users, as in the case of Local Area Network (LAN) interconnection. Although the data cell arrival process in ATM networks has not yet been identified, actual data packet arrival processes have been investigated.

It is well known that generation of data from a single data source is well represented by a Poisson arrival process (continues time) or by a geometric inter-arrival process (discrete time). When information loss occurs, these kinds of services use retransmission as a way of recovering information. The retransmission of the complete data frame is executed every time there is cell loss.

2.4.3.1 Interactive Data Transmission

A single packet is generated at each time. This could be either a fixed length or a variable length packet. The length of the packet is represented by a certain distribution of fixed mean.

This traffic is of bursty nature, relatively short in length, and requires relatively small delay in transmission. Delay variance is not a major problem, but error free transmission is an important requirement.

Examples of such traffic are transaction/credit card verification, hotel/airline reservation, WWW access and various short message transmissions.

2.4.3.2 Bulk Data Transmission

The nature of the traffic is similar to the earlier case, but now messages consist of a number of packets. This is a batch arrival case and the arrivals of the packets that make up the message are not independent. Since ATM networks have a fixed cell size, it may happen that a data packet of either variable or fixed size is fragmented into several cells.

The performance requirements are similar to the previous case, but a slightly higher average delay might be acceptable. Examples for bulk data transmission are file transfer, database information acquisition etc. Candidates to model data sources are the two state MMPP, also called Switched Poisson Process (SPP), including the Interrupted Poisson Process (IPP) and the Geometrically Modulated Deterministic Process (GMDP), including the on–off model.

2.4.4 Multimedia

The term multimedia is used to refer to the representation, storage, retrieval, and transmission of multiple media, such as text, voice, graphics, image, audio and video. Multimedia applications constitute a significant future market. Examples include teleconferencing, entertainment video, medical imaging, distance education, telemarketing and advertising. Each of these applications consists of two or more information types, which are listed above. It has to be noted that a strong correlation between successive cell arrivals is characteristic of many multimedia traffic sources.

Traffic models of information types (i.e. video, voice, data), which are put together in multimedia services are presented throughout this section. The extension of these models to characterise the integrated environment of a multimedia service is an important task and is currently under extensive study [28]. Despite this, the MMPP is widely used to model superposed traffic of different information types, and could therefore be used to model multimedia traffic.

2.5 Influences on Traffic Characteristics

The fundamental difficulty in identifying traffic characteristics is the interdependence of the traffic flow on the network itself.

Care must be taken in traffic modelling since every layer in the OSI reference model (Figure 2.1) is a function of higher and lower layer activity and is influenced by the network behaviour or human interaction with the application layer. For example traffic characteristics of call duration, bandwidth, burstiness, burst duration, peak and sustained packet rate will vary for the same generic application (e.g. video) with the quality of service that the customer has chosen.

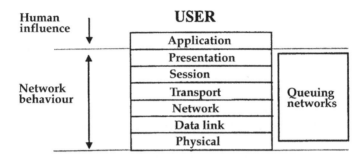

Figure 2.1   Influences on different protocol layers

Video applications, for example, already appear in a vast array of flavours based on proprietary and international standards. The traffic characteristics and service data will depend on human interaction, i.e. whether motion is active or passive, and technical preferences such as the resolution, colour or greyscale etc.

The video and audio codec employed impacts the traffic characteristics dependant on what access network the codec has been optimised for e.g. ISDN, POTS, ATM, LAN, mobile etc. To complicate the issue further new standards on codec technology are continually emerging and developing. Protocols at each layer of the OSI reference model introduce variations in flows:

- At the session layer, ubiquitous protocols such as HTTP 1.0 which dominates the data traffic flows on the Internet are likely to have been superseded by a more recent version, HTTP 1.1. Mobile specific protocols such as Uplink and languages such as WML (Wireless Application Protocol Mark-up Language) will alter the format and method of data transfer.
- Transmission control protocols e.g. TCP are maturing and protocol extensions are being added that will dramatically alter the measured packet flow. Experiments have shown [31] that the TCP congestion control algorithm had more control over the loss rate in the shaping buffer than the service rate of the buffer. TCP responds to corruption as if there is loss of information bits or congestion in the network and responds accordingly. Successive lost segments triggers congestion avoidance which continually resets the slow start procedure which leads to an inefficient use of resources over a satellite link with significant propagation delay. Alternatively any loss under steady-state operation is interpreted as

congestion and congestion control response is invoked. In either case the steady-state traffic characteristics will be modified.

• The current Internet Protocol standard, which underlies the operation of the Internet, will possibly be superseded by IPv6 (or later) and other network layers such as IPX and SNA must be catered for. The relative dominance of these network protocols is unknown at this time.

Traffic shaping at the network edge would alter the characteristics of the traffic entering and leaving the network e.g. through the addition of delay, smoothing of burstiness for non-real time traffic and by cell prioritisation.

2.6 Source Models

There are usually, several alternatives to represent a particular traffic source and different levels of complexity. The most important criteria upon which the selection was based is briefly discussed. First of all, the chosen source model must be accurate with respect to our assumptions. It should be close to reality and the different parameters should not have only a statistical but also a physical meaning. Analytical and simulation results of the model should be compared, if possible, to measured performance of real traffic sources to verify the source models accuracy and validity.

From the analytical point of view, tractability (superposition/queuing) is an important feature. This means that the use of the source models in analysis should lead to solutions that lend themselves for numerical computation. In many cases, general methods such as iterations to solve systems of linear equations, aggregation methods to reduce the dimensionality, matrix analytical methods to solve structured Markov chains etc. could be applied. Often, the exploitation of the special structure of the processes involved, may make the model much more suitable for numerical solutions, without losing its probability interpretation.

Another important feature of source models is its generality and usability. A typical model should be able to represent a large class of sources with similar characteristics. Since most of the source models are also used in simulations, care should be taken so that it is possible to represent the model in a simulation environment. It is also important that the model is statistically stable, otherwise there may be significant problems in the overall network simulation model that might be difficult to detect. The statistical stability is measured over a period of time, which is proportional to the highest level of resolution in time specified by the source model and its number of different states.

Finally the number of parameters of the model should be taken into account. This number is usually directly related to the complexity of the description of the model. The aim is to use a model that is adequate for our purpose, but uses a limited number of parameters. This makes the analysis easier and the computation faster.

2.6.1 Multi-State Markov Source Models

2.6.1.1 General Modulated Deterministic Process (GMDP) Model

The GMDP is based on a finite state machine having n states. In each state, cells are generated with constant interarrival time T_i (therefore it is called a deterministic process), the index i being the state. The number of cells which are emitted in state i may have a general discrete distribution. Usually, the GMDP also includes silence states where no cells are generated and the duration of these states may also have a general discrete distribution. If the burst and silence duration have an exponential distribution then the model is called a Markov Modulated Deterministic Process (MMDP). The on–off model is a two state MMDP with one silence state.

The state transitions are governed by a transition matrix where each element denotes the probability of moving from state i to state j once the sojourn period expires. Usually, voice traffic sources can be characterised when using this model with 2 states, whilst video traffic sources may need 3 states to be characterised.

2.6.1.2 The On–Off Model

One traffic model, which is widely used for the characterisation of ATM sources, is the on–off source model. This model has been successfully used to realistically model packetised speech, still picture and interactive data services. According to the on–off model the ATM cell stream from a single source is modelled as a sequence of alternating burst periods and silence periods. This model is a 2-state Markovian representation of an ATM source as shown in Figure 2.2. The duration of each burst is exponentially distributed with mean $1/a$ ms. During such a period ATM cells are emitted with *constant* interarrival time T ms, where $T = 1/PCR$. After generation of the ATM cells an exponentially distributed silence period with mean value $1/b$ ms follows. This corresponds to a geometrically distributed number of packets per active period (i.e. burst), with mean value $1/(aT)$, followed by an exponentially distributed silence period, with mean value $1/b$.

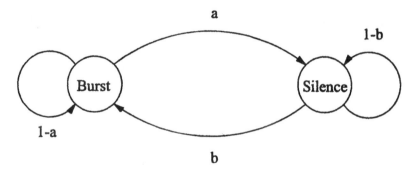

Figure 2.2   Two-state Markovian representation of an ATM source

Note that this model is a special case of the GMDP; it is equivalent to a two state MMDP with one silent state.

The on–off traffic source model as shown in Figure 2.3, can be described by the parameters (PCR, m, β, t_{on}) as follows:

$$PCR = 1/T,\ t_{on} = a^{-1},\ m = a^{-1}/T(a^{-1} + b^{-1})\ \text{and}\ \beta = (a^{-1} + b^{-1})/a^{-1} \qquad (2.4)$$

where a and b are the transition rates, i.e. a is the inverse of the mean burst duration, b is the inverse of the mean silence duration, m is the mean cell rate, β is the burstiness and t_{on} is the average burst duration.

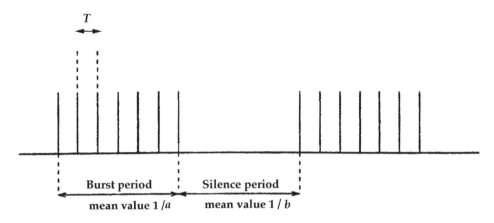

Figure 2.3   The on–off source model

2.6.1.3 The Interrupted Poisson Process (IPP)

The IPP is a Poisson process that is alternatively turned on for an exponentially distributed period of time (active period where cells are emitted) and turned off for another exponentially distributed period of time (silent period), like the on–off model. The difference however is that during the active periods, the interarrival times of cells are *exponentially* distributed (i.e. in a Poisson manner).

The advantage of modelling the arrival process from a single voice source as IPP is that the aggregated arrival process from multiple sources can be modelled by a MMPP. This is due to the fact that the IPP is a special case of an MMPP and the superposition of MMPPs is an MMPP.

Let r_1, r_2 and λ respectively denote the average duration of the active and silent periods, and the cell generation rate during the active period. The simplest way to determine these IPP parameters is to set the mean talkspurt length ($a^{-1} = t_{on}$) to the mean sojourn time of the cell arrival process r_1, to set the mean silence period length [$b^{-1} = t_{on}(\beta - 1)$] to r_2 and to set the mean cell generation rate during a talkspurt λ^{-1} to that of the cell arrival process (T). But this parameter matching underestimates the performance, therefore the two-moments and peakedness method has been proposed [29]. Let m, c and z be the mean arrival rate, squared coefficient of variation, and peakedness of the cell arrival process. Under the assumption of exponential talkspurt and silence distribution, they are given by the following equations:

$$m = \frac{a^{-1}}{(a^{-1} + b^{-1})T} \tag{2.5}$$

$$c_a^2 = \frac{1 - (1 - aT)^2}{(a + b)^2 T^2} \tag{2.6}$$

$$z = \left[1 - \left(aT + \frac{abT}{m^{-1} + b}\right)e^{-T/m}\right]^{-1} - \frac{mb}{(a + b)T} \tag{2.7}$$

Then the parameters in an IPP can be expressed as:

$$\lambda = \frac{1}{m} + \frac{(c^2 - 1)(z - 1)m^{-1}}{c^2 + 1 - 2z} \tag{2.8}$$

$$r_1 = \frac{2(z - 1)m^{-1}}{(z - 1)(c^2 - 1) + c^2 + 1 - 2z} \tag{2.9}$$

$$r_2 = \frac{2(c^2 - 1)(z - 1)^2 m^{-1}}{[(c^2 - 1)(z - 1) + c^2 + 1 - 2z](c^2 + 1 - 2z)} \tag{2.10}$$

2.6.1.4 Interrupted Bernoulli Process (IBP)

The IBP is a discrete version of the IPP. Time is slotted, with a slot length being equal to the cell in the medium. A slot is either in an active state or in a silent state. A slot in an active state contains a cell with probability a and no cell with probability $(1 - a)$, while no cell arrive in a silent state. Given that the slot is in the active state (independent of whether the slot contains a cell or not), the next slot is also in the active state with probability p and changes to the silent state with probability $(1 - p)$. Similarly, given that the slot is in the silent state, the next slot is also in the silent state with probability q and changes to an active state with probability $(1 - q)$. Accordingly, both the active period, $\Pr(X = x)$, and the silent period, $\Pr(Y = y)$ are geometrically distributed. That is

$$\Pr(X = x) = (1 - p)px^{-1}; \Pr(Y = y) = (1 - q)q^{y-1} \qquad x, y \geq 1 \tag{2.11}$$

with respective average duration times equal to $1/(1 - p)$ and $1/(1 - q)$.

2.6.1.5 The Birth–Death Process

A birth–death process is a Markov model where only transitions to adjacent states are generated. The continuous-time birth–death process is used to model voice and video. This process can be viewed as the superposition of N independent homogeneous on–off sources. This continuous-time process is a fluid approximation model and bit rates can be seen as switching between states with discrete values, and the time spent in each state is given by a random Poisson time sequence.

For voice, instead of modelling the individual information sources, the total bit rate of N independent active voice sources is modelled. To model the actual video source, bit rate takes on only discrete quantised values and are assumed to sampled at random Poisson times in the time domain. If $p(i,j)$ is the transition rate from state i to j, the birth and death rates are given [30] by:

$$p(i, i + 1) = (N - i)b \qquad i < N \tag{2.12}$$

$$p(i, i - 1) = i \times a \qquad i > 0 \tag{2.13}$$

where a and b are the transition probabilities.

The equilibrium probability of being in state i is given by the binomial distribution:

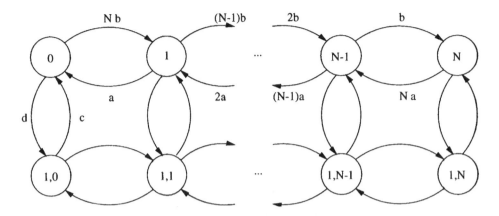

Figure 2.4   Two-dimensional birth–death process

$$P_{\text{Bi}} = \binom{N}{i} \beta^i (1 - \beta)^{N-i} \qquad \beta = b/(a + b) \tag{2.14}$$

The two dimensional, continuous time birth–death process shown in Figure 2.4 can be used to model jumps to higher or lower bit rates in video scene changes. Each dimension of the model can be viewed as the one-dimensional birth–death process discussed above. c and d are the transition probabilities to low-activity and high-activity levels respectively. Note that the cell rate in state N is NA_1 and is $A_h + NA_1$ in state $(1, N)$.

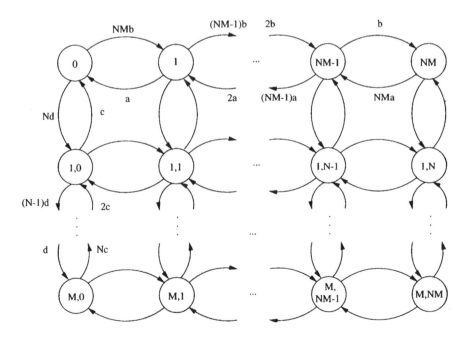

Figure 2.5   State-transmission-rate diagram for aggregate source model

If a single video source is modelled in this manner, the bit rate when multiple information sources are multiplexed can be modelled with the same structure. Thus the multiplexing of N video sources can be modelled with a state-transition-rate diagram like that shown in Figure 2.5.

2.6.1.6 The MMPP

The MMPP has been extensively used to model various B-ISDN sources, such as voice, video, as well as characterising superposed traffic. It has the property of capturing both the time-varying arrival rates and correlation's between the interarrival times. Also, if individual traffic sources are modelled by an MMPP, the superposition of different sources can be described by an MMPP.

An MMPP is a doubly stochastic Poisson process. The arrivals occur in a Poisson manner with a rate that varies according to a n-state (phase) Markov chain, which is independent of the arrival process.

As the simplest case, Figure 2.6 shows the 2-state MMPP (also called Switched Poisson Process (SPP)) having Poisson arrival rate λ_j in phase j, $j = 1,2$, which appears alternately exponentially distributed sojourn time with mean r_j^{-1}. This is characterised by (R, Λ) where R is the infinitesimal generator of the underlying Markov chain and Λ the arrival rate matrix, defined by:

$$R = \begin{bmatrix} -r_1 & r_1 \\ r_2 & -r_2 \end{bmatrix} \qquad \Lambda = \begin{bmatrix} \lambda_1 & 0 \\ 0 & \lambda_2 \end{bmatrix} \tag{2.15}$$

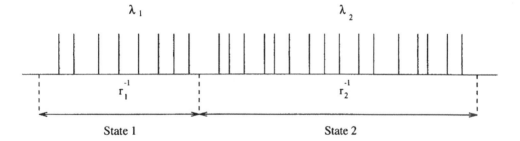

Figure 2.6　Two-state MMPP

The n-state MMPP is similarly characterised by (R, Λ) with each matrix of $n \times n$ size. In special cases, the MMPP becomes a renewal process, which is characterised by statistically independent and identically distributed interarrival times. If $\lambda_1 = \lambda_2 = \lambda$, the MMPP reduces to a Poisson process with rate λ. If $\lambda_2 = 0$, it is called an IPP.

2.6.1.7 Superposition of MMPPs

The superposition of MMPPs is also an MMPP. Therefore it can be used to model superposed heterogeneous traffic. Consider N MMPP models, each with parameters R_i and Λ_i. Then, the transition rate matrix R and the arrival rate matrix Λ of the superposed process are:

$$R = R_1 \oplus R_2 \oplus ... \oplus R_N \qquad \Lambda = \Lambda_1 \oplus \Lambda_2 \oplus ... \oplus \Lambda_N \tag{2.16}$$

where $\oplus$ denotes the Kronecker sum defined below. We note that both R and Λ are $k \times k$ matrices, where $k = 1,...,N$.

The Kronecker sum of two matrices R_1 and R_2 is defined as:

$$R_1 \oplus R_2 = (R_1 \oplus I_{R_1}) + (I_{R_2} \oplus R_2) \tag{2.17}$$

where I_{Ri}, $i = 1,2$ is an identity matrix of the same order as matrix R_i and x denotes the Kronecker product, which is defined for two $C = \{c_{ij}\}$ and $D = \{d_{ij}\}$ as:

$$C \oplus D = \begin{bmatrix} c_{11}D & c_{12}D & ... & c_{1m}D \\ ... & ... & ... & ... \\ c_{n1}D & c_{n2}D & ... & c_{nm}D \end{bmatrix} \tag{2.18}$$

As the number of the superposed processes increases, the number of states of the MMPP increases and it becomes very complex to solve queues with a large number of arrival streams. Due to the complexity of matrix analysis with a high number of states it is difficult to obtain the source model of multiplexed traffic sources. To reduce complexity, the super-posed process may be approximated by a simpler process that captures important character-istics of the original process as closely as possible. The simplest model that has the potential to approximate an MMPP with a large number of states accurately, is the two phase MMPP, which was described above.

2.6.2 Self Similar Models

Recent analysis of real traffic traces [1,2] clearly show that Internet traffic exhibits self-similarity features. Precisely, let X_n be the number of packets arriving in interval n and form the aggregated process $X_n(m)$ consisting of the sample mean of m non-overlapping intervals. The measured process X_n is self-similar since $X_n(m)$ is equal to X_n in distributional sense. Therefore, the process looks like a fractal: no matter the time scale that we consider the distribution remains invariant. On the other hand, we note that the burstiness of the packet arrival process also remains invariant with increasing time scale, thus severely affecting network performance.

Self-similar traffic exhibits long-range dependence in the packet counting process in contrast to Poissonian models, which show independent increments. A stationary stochastic process is long-range dependent if the autocorrelation function is non-summable. Regarding the process X_n it turns out that the autocorrelation function decays slowly as a power law $(r(k) \approx H(2H - 1)k^{(2-2H)})$. The parameter $H(0.5 < H < 1)$ is called the Hurst parameter and serves to the purpose of measuring the long range dependence of X_n. The effect of such slowly decaying correlation in the queuing performance is rather striking, if we compare to the performance figures obtained with Poissonian input.

Self-similar models can be grouped in the following categories: Fluids such as the Frac-tional Brownian Motion, Point Processes such as Fractal Renewal Processes [10,11] and Deterministic Transformations such as chaotic maps. Point processes are used to model Internet services such as the virtual terminal (Telnet), in which no bulk traffic is generated and thus differ from fluid-flow behaviour. Deterministic transformations serve to the purpose of modelling the effect of some deterministic features such as the TCP timers that explain part of the self-similarity phenomena.

However, it has been shown [2,4] that Internet traffic is dominated by TCP transactions (mostly from the FTP and WWW), for which a fluid-flow approach applies. Such transactions are generated by a large number of independent users, so that the arrival process is Poisson. However, Poisson arriving heavy-tailed bursts tend to a Fractional Brownian Motion in the limit, which is a well-known self-similar process. A heavy-tailed distribution shows finite or infinite mean depending on shape parameter, and infinite variance. Recent measurements show that the size distribution of WWW pages in the Internet is heavy-tailed (Pareto distribution) [3], thus providing a phenomenological explanation of self-similarity in terms of file sizes and transmission duration.

2.7 Geographic Traffic Models

For the end-to-end performance analysis of a satellite communication system it is more appropriate to consider aggregate traffic modelling, and especially geographic traffic modelling, for two reasons: First, satellite network planning concerns with cells of big service coverage so it is not possible to realise models in which mobile subscribers are modelled separately, and second, its performance is strongly dependent on the very non-homogeneous geographic characteristics of the population. Thus, a global geographic traffic model is needed which will predict the spatial and temporal distribution of the traffic intensity offered by any studied earth segment.

These models characterise the spatial and temporal distribution of the traffic intensity as a function of various demographic and cultural data (population, income, mobile penetration) [5–7]. Geographic traffic models can be *service dependent*, i.e. different for each kind of offered service (adequate for the service) and *adaptive*, i.e. the earth splitting procedure is continuously adapted to traffic loading of the examined region.

Some geographic traffic models have already been presented in the literature [5,6,9] but the detailed description of load input is still lacking. Usually landmasses are divided into big areas over which the estimated traffic is distributed uniformly [6]. Some other proposed distributions are the Gaussian, the triangle, Erlangian or generalised distribution [7,8]. In Ref. [7] the non-uniform distribution of users is examined together with its impact on the network performance.

2.7.1 Traffic Intensity – Forecasting Techniques

Traditional forecasting methods were based on regression [12] and recently on fuzzy and neural networks [18] and expert systems. Recently diffusion models [14,15] have been considered more promising for satellite networks since no history is available for many of the emerging services. A specific diffusion model, called Generalised Bass Model (GBM) is presented in [16] for the estimation of market growth for emerging multimedia services. The paper describes the procedure service modelling first and then it proceeds with the distribution of traffic demand in each cell.

The central problem with forecasting demand for broadband telecommunications services is that there is so little historical data on which to base them. The applications which will place demands on these networks have, on the whole, yet to be developed, or are only now being launched. However, the telecommunications industry is slower to change than the computer industry and needs to think further ahead if it is to provide the necessary communications infrastructure.

References

[1] W.E. Leland, M.S. Taqqu, W. Willinger and D.V. Wilson, 'On the self-similar nature of ethernet traffic', *IEEE/ACM Transactions on Networking*, Vol. 2, No. 1, pp. 1–15, 1994.

[2] V. Paxson and S. Floyd, 'Wide-area traffic: the failure of Poisson modelling' *IEEE/ACM Transactions on Networking*, Vol. 3, pp. 226–244, 1995.

[3] M. Crovella and A. Bestavros, 'Self-similarity in World Wide Web traffic: evidence and possible causes', *IEEE/ACM Transactions on Networking*, Vol. 5, No. 6, pp. 835–845, 1997.

[4] J. Aracil, 'On Internet Traffic Self-Similarity', COST253 TD (98) 002.

[5] G. Schorcht, 'A global traffic model for simulation of the network load in mobile satellite communication Networks' *Proceedings of ICT98*, Greece.

[6] M. Werner, A. Jahn, E. Lutz and A. Bottcher, 'Analysis of system parameters for LEO/ICO–satellite communication networks' *IEEE Journal in Selected Areas in Communications* February 1995.

[7] A. Jamalipour, M. Katayama, and A. Ogawa, 'Traffic characteristics of LEOS-based global personal communications networks' *Communications Magazine*, February 1997.

[8] I. Norros, 'Traffic Aspects of Data Services Over Wideband CDMA Third Generation Mobile Communications' COST257 TD (99) 13.

[9] K. Tutschku and P. Tran-Gia., 'Traffic Estimation and Characterization for the Design of Mobile Communication Networks', COST256 TD (97) 47.

[10] P. Mannersalo and I Norros, 'Multifractal Analysis: a Potential Tool for Characterising Teletraffic', COST257 TD(97)32.

[11] R.H. Riedi, et al., 'LAN Traffic is Multifractal: a Numerical Study', 1996, http://www-syntim.inria.fr/fractales/.

[12] R. Hu, 'Development and Analysis of ABR Congestion Control Techniques in Wide Area ATM Networks ' PhD. University of Kansas, 1998.

[13] A. Abaye, 'Power Systems Load Forecasting by Neural Networks', PhD. SMU Dallas, 1995.

[14] M. Hopkins, 'A Multifaceted approach to forecasting broadband demand and traffic' *IEEE Communications Magazine*, pp. 36–42, February 1995.

[15] M. Lyons, et al., 'Dynamic modelling of present and future service demand ' *Proceedings of the IEEE*, Vol. 85, No. 10, 1997.

[16] A. Abaye, et al., 'Forecasting methodology and traffic estimation for satellite multimedia services', *ICC '99*.

[17] ITU-T. Rec. I.371, 'Traffic and Congestion Control in B-ISDN', May 1996.

[18] ATM-Forum, 'Traffic Management Specification Version 4.0'. Technical Report 0056, ATM-Forum, April 1996.

[19] CCITT. Rec. G.711, 'Pulse Code Modulation (PCM) of voice frequencies', October 1984.

[20] K. Sriram and W. Whitt, 'characterising superposition arrival process in packet multiplexers for voice and data', *IEEE Journal in Selected Areas in Communication*, Vol. 4, No. 6, pp. 833–846, 1986.

[21] P.T. Brady, 'A model for generating on/off speech patterns in two way conversations', *Bell Systems Technical Journal*, Vol. 48, pp. 2445–2472, 1969.

[22] RACE CATALYST R2074, 'Intermediate modelling and simulation report', University of Surrey, 1993.

[23] N. Ohta, 'Packet Video: Modelling and Signal Processing', Artech House, 1994.

[24] M. Nomura, T. Fujii and N. Ohta, 'Basic characteristics of variable rate video coding in ATM environment', *IEEE Journal in Selected Areas in Communication*, May 1989.

[25] B. Maglaris, D. Anastassiou, P. Sen, G. Karlsson and J.D. Robbins, 'Performance models of statistical multiplexing in packet video communication', *IEEE Transactions on Communications*, Vol. 36, No. 7, pp. 834–843, July 1988.

[26] R. Grunenfelder, J. Cosmos, S. Manthorpe and A. Odinma-Okafor, 'Characterisation of video codecs as autoregressive moving average processes and related queuing system performance', *IEEE Journal in Selected Areas in Communications*, March 1991.

[27] P. Sen, P.B. Maglaris, N. Rikli and D. Anastassiou, 'Models for packet switching of VBR video sources', *IEEE Journal in Selected Areas in Communication*, Vol. 7, No. 5, pp. 865–869, 1989.

[28] T. Ors, P. Taaghol and R. Tafazolli, 'Initial Traffic Modelling Report', MVCE/SRU/WPS03/C01/001, 26th November 1997.

[29] I. Ide, 'Superposition of interrupted Poisson process and its application to packetized voice multiplexer', IEE 12th International Teletraffic Congress, 1988.

[30] R. Guerin and J. Heinanen, 'UBR+ Service Category Definition', ATM-Forum 96-1598.

[31] A. Manthrope, 'The Danger of Data Traffic Models', COST242 Technical Document 95–25, Stockholm, 30th April 1995.

3

Transmission Schemes[☆]

In the future, radio communication systems will offer different services through single user equipment; furthermore, it is expected that different radio communication systems will merge into one universal communication system. Today, there are two known concepts for providing wireless communication services: terrestrially based communication systems and satellite communication systems. Each concept has its specific characteristics, and consequently, advantages and disadvantages. The main disadvantages of terrestrially based systems are:

- Poor radio channel quality; Rayleigh fading limits the cell radius and transmitted data rate and
- Terrain shadowing causes gaps in radio coverage.

On the other hand, signal propagation delay in satellite systems causes impairments in voice communications, especially in GEO constellations and indoor coverage is generally poor. The high power handsets used in satellite systems cause health concerns. However, a third option exists, High-Altitude Aeronautical Platform (HAAP) [13], which has the major advantages of both concepts and avoids many of the pitfalls of satellite and terrestrial systems. Aeronautical platforms are flying between 17 and 22 km above the Earth surface; therefore the propagation delay and the handset terminal power emission are similar to terrestrial systems. The radio channel propagation is a free space like channel at distances comparable to terrestrial systems, with shadowing of the radio signal as in satellite systems. Indoor coverage also is possible with HAAP systems. A few studies have been carried out analysing the HAAP platforms as a GSM base station, and a number of papers have been published about the capabilities of HAAP as a local multi-point distribution systems. On the other hand, some serious disadvantages of the high altitude aeronautical platforms exist: the power consumption of the base station at high altitude aeronautical platforms is limited as in satellite systems; and there is public concern about a potential crash of the large unmanned aeronautical vehicles.

It is obvious that terrestrial systems alone are not sufficient for global Earth coverage, and the non-terrestrial systems have to be added to cover gaps in terrestrial system radio coverage and to introduce the new broadband services.

Section 3.1 is dedicated to the modulation schemes, and in particular we present an efficient modulation scheme for satellite communication, the Continuous Phase Modulation

---
[☆] Part of the work presented in this chapter has been carried out by Bertrand Ficini and Marius Bjerke during their trainee periods at Telenor R&D in 1997 and 1998, respectively.

(CPM). The radio interface of the future non-terrestrial systems should be capable of adapting to a wide range of data rates, traffic densities and propagation conditions. Therefore, the adaptive modulation techniques, capable of adjusting to different channel conditions, available energy on the platform and required data rates, should be used.

In Section 3.1.2, we describe one possible solution for a variable rate modulation scheme suitable for implementation in communication systems where the power consumption is limited and high data rates are required – variable N-Minimum Shift Keying (MSK).

Section 3.2 presents the channel code, and in particular a coding scheme is considered appropriate for a certain type of communication. The convergence of broadcasting and communications means that both meshed networks and multi- and unicast star networks should be considered. Today, the most commonly proposed transmission scheme is packet switching, and the applications are all in one network, designated as multimedia. The multimedia application, together with the routing and the high quality requirements, represent quite a challenge in satellite communications. One of the sections in the network where this challenge must be met, is the physical layer implementation, where the modulation and channel coding have to face stringent quality demands. For packet switched traffic and high quality applications, Bit Error Rate BER in the order of 10^{-8} to 10^{-10} are wanted for Signal-to-Noise Ratio (SNR) of a few decibels (dB) for spectral efficiencies above 1 bit/s per Hz. We propose the use of a Turbo code, particularly suited for transmission of short independent packets, the Turbo-Decoded Product Codes (TPC).

In Section 3.3, we discuss some of the problems related to synchronisation, and in particular techniques for Döppler shift compensation. In the network type we consider in this COST action, there are a number of different levels of synchronisation; in addition to the classical frequency and symbol clock synchronisation, we have frame/packet synchronisation, code synchronisation, Döppler frequency shift synchronisation, and timing adjustments due to the motions of both the terminal and the satellites. In our system, the mobility of the terminal is considered zero, since we are primarily concerned with inter-LAN connections through mobile satellites. Even if some kind of portability of the terminal is possible, the motion would be negligible compared with the satellite motion. The Döppler frequency shift synchronisation is unique and may be handled in many ways. The uniqueness lies in the fact that the Döppler shift is a deterministic feature of the system as soon as the system characteristics (geometry and relative speed) are known. As a result of this, it is possible to correct the Döppler frequency shift at the transmitter, or more conveniently in our case, in the satellite. However, this is not without problems, the compensation in the satellite will lead to increased satellite complexity. Hence, we have studied two algorithms capable of correcting the Döppler shift at the receiver as if the magnitude of the shift was unknown, in order to determine the effect both on performance and complexity.

Finally, Section 3.4 highlights the relevance of introducing users' random spatial location into models of satellite telecommunication systems. Our illustration shows the computation of the first moments of the cumulated interference power at a satellite in a LEO satellite system where transmission is based on the Code Division Multiple Access (CDMA) principle. The traffic distribution over a geographical area is critical in determining the total interference power for the uplink. The received power is directly related to the exact location of the source of interference. The Isotropic Phase-Type Planar Point Process (IPhP3) is a spatial point process that provides a tractable tool to model user location in the area of interest. Latouche and Ramaswami first defined the process in [19], and it has been extensively studied since (see Remiche [20] for references). In this case, a mark is attached to each point of the process (each source of interference). This mark consists of a deterministic function of the source location itself. In this study, the associated mark is the value of the received interference power at a given satellite. The model used is directly inspired from Jamalipour et al. in [18]. A spatial dimension to the model by characterising the random location of the interference source is also included.

3.1 Modulation Techniques

When considering modulation techniques for packet-switched traffic over satellites, both traffic type and channel impairments must be considered. The most severe restrictions stem from the satellite channel, which is band-limited, noisy, non-linear and may present fading characteristics in certain frequency bands. The ideal modulation would be bandwidth-efficient and resistant to noise, Inter-Symbol Interference (ISI), Adjacent Channel Interference (ACI), and non-linearities. Unfortunately, there are inconsistencies and one must be traded-off against the other.

   We will briefly discuss the benefits and drawbacks of a selection of modulation schemes – comprising Quaternary Phase Shift Keying (QPSK), Differential QPSK (DQPSK), Quadrature Amplitude Modulation (QAM) and CPM. The discussion will be based BER performance, spectral efficiency, spectra, robustness vs. non-linearities and complexity.

3.1.1 Overview of Modulation Schemes for Satellite Transmission

3.1.1.1 QPSK
The QPSK is a simple memoryless modulation conveying 2 bits per symbol. ISI-free transmission on Additive White Gaussian Noise (AWGN) channel is usually obtained with Nyquist filtering. In that case, the spectrum will occupy a bandwidth of $(1 + \alpha)/T_s$, where α is the Nyquist filter roll-off factor and T_s the symbol period. QPSK is not a constant envelope modulation, hence High Power Amplifier (HPA) non-linearities combined with the digital filtering will induce both AM/AM and AM/PM distortion, as well as spectral regrowth. AM/PM distortion may be corrected with a simple Automatic Gain Control (AGC), but AM/AM distortion and spectral regrowth will exist and lead to ISI and ACI, respectively (see Figure 3.1). The solution is to make sure of a sufficient back-off of the operating point vs. the saturation point of the HPA. QPSK is a constellation with points on a

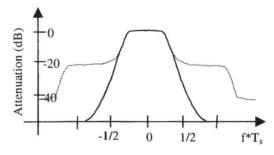

Figure 3.1   Ideal QPSK spectrum and spectral regrowth of a QPSK signal with 1 dB back-off and a roll-off $\alpha = 0.35$

circle and makes the modulation quite resistant to non-linear effects. Usually 0.5–1 dB back-off is sufficient for reasonable performance requirements with adequate Nyquist filter roll-off factors. The lower the roll-off factor, the more spectral regrowth and the more memory effects will be seen on the constellation points, resulting in increased ISI. As mentioned above, QPSK is considered a simple modulation scheme, and combined with the relatively good performance of a satellite channel, this modulation is the most popular modulation scheme for satellite communication. Limited spectral efficiency is the price to pay for robustness and simplicity.

DQPSK

Differential QPSK is a QPSK modulation with memory, chosen in situations where proper estimation (knowledge) of the carrier phase is difficult to obtain [15]. In such cases, it is possible to solve the problem by employing the robust DQPSK modulation and coherent or non-coherent demodulation, at the expense of performance degradation.

In coherent demodulation, the phase is recovered with a $k\pi/2, k = 1, 2, 3$, ambiguity which is solved by simple binary logic. The SNR penalty in this case is quite small, 0.2 dB at a BER of 10^{-6}, when compared to the QPSK modulation. This penalty is induced by the doubling of the binary errors at the receiver due to the consideration of two consecutive symbols. The frequently used binary differential detection requires minimal additional circuitry and minimal SNR penalty, and has become an inherent function of the QPSK modulation scheme.

The non-coherent demodulation will operate without any phase information at all. In that case the operation on two consecutive noise symbols involves a multiplication of the symbols with a much higher impact on performance. The SNR penalty is about 2.3 dB when compared to coherent QPSK demodulation. Simplicity of the demodulator implementation is therefore achieved.

In any case, non-linear effects remain identical to QPSK modulation, as well as the spectral efficiency and the spectrum.

3.1.1.2 QAM

QAM is a memoryless modulation with constellation points forming a quadratic- or cross-formed grid in the I–Q plane. The idea is to stay as circular as possible, while maintaining symmetry with respect to the I and Q axis as well as the diagonals. For an M-QAM the number of constellation points is M. QAM can be considered as being within the same modulation family as QPSK because a QPSK also forms a 4-QAM. This is why many of the QAM characteristics can apply to the QPSK. A doubling of M leads to an increase of 1 bit/ T_s in spectral efficiency, and an additional 3 dB in SNR for the same BER performance.

The spectral properties are identical to QPSK on an AWGN channel. The main difference between M-QAM and QPSK is that the constellation points of the QAM are no longer situated on one circle, causing the vulnerability towards non-linearities to increase (see Figure 3.2). Back-off values which are necessary for maintaining equivalent BER, increases rapidly by several dB when doubling M if no countermeasures are taken. Reduction of non-linear effects has been studied for some years, and many interesting techniques have been proposed, mainly predistortion techniques [1]. The complex techniques and comfortable spectrum allocations on satellites, have limited the interest in prototype implementation. Spectral efficiency, which is also important in satellite communication and technological

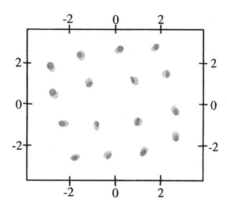

Figure 3.2   Typical 16-QAM constellation at non-linear channel output with AM/AM and AM/PM effects (without noise). The clustering of the constellation points is due to interaction between non-linearities and the memory of the Nyquist filtering

development, reduces complexity and cost. These techniques will become highly relevant for practical implementation in the future.

One solution for increasing the spectral efficiency while maintaining the signal constellation on a circle, is to use M-PSK. This solution will to some extent prevent the degradation due to non-linear effects of the HPA. However, the increase in signal points seldom goes beyond $M = 16$ due to the increase in decoder complexity and reduction in Euclidean distance between signal points.

3.1.1.3 CPM

CPM is a constant envelope modulation with memory [2]. The information is contained in the phase transitions from one symbol sample at nT_s to another symbol sample at $(n + k)T_s$, $k = 1,2,...$ CPM is a modulation family classifying types according to the memory involved and the pulse shaping. We can mention:

- LRC – raised cosine with pulse length L
- LREC – rectangular frequency pulse of length L
- LSRC – spectral raised cosine of length L
- TFM – Tamed Frequency Modulation
- GMSK – Gaussian-shaped MSK

Within this classification, some of the most well-known schemes are unique to classes, i.e. MSK is a 1REC with modulation index $h = 1/2$. The difference in performance and spectral efficiency is quite important when comparing one modulation with another. However, some tendencies may be mentioned. The BER performance in an AWGN channel depends on the minimum distance between neighbour signal sequences. This distance will globally increase with the modulation index, h, and with increasing pulse duration, L. The increase of the minimum distance leads to a better BER performance. The plots in [2] show that the minimum distance as a function of the modulation index is not a strictly increasing function – there are values of h for which the minimum distance suddenly decreases. The corresponding values of h are called weak indexes.

CPMs are ideally of infinite bandwidth. Due to their non-linear nature, optimal linear filtering at the receiver is not possible, and their spectral efficiency is therefore difficult to compare with linear schemes such as QPSK. As a rule, the spectrum of CPM is usually expressed as a 10, 20 or 30 dB spectrum, defining that 90, 99 or 99.9% of the signal energy is within the spectral limits, respectively. In general, CPM are more spectral efficient than linear modulation with the same M (with the same number of bits/T_s), and that the spectral-efficiency increases with M and/or L. The temporal transitions in CPM are smoother than for the linear modulation schemes, which leads to lower sidelobes. Increasing M, L or decreasing h, all lead to reduced high-frequent energy. Figure 3.3 shows a trellis representation of a CPM signal with $h = 2/3$, $M = 2$ and $L = 3$ from [2].

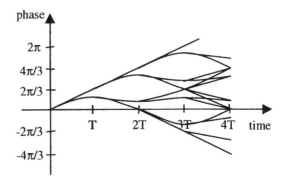

Figure 3.3   Phase tree of a CPM with $M = 2$, $L = 3$ and $h = 2/3$ [3]. The pulse is an RC

The main advantage of CPM in satellite communication is that the constant envelope makes the signal invulnerable towards non-linear distortions. This means that no precaution is needed when operating the HPA; it is possible to profit from the peak transmitter power at saturation without signal degradation.

On the other hand, the main disadvantage of CPM schemes is their demodulator complexity. The CPM being a non-linear modulation with memory leads to a Maximum Likelihood Sequence Estimation (MLSE) with a high number of adapted filters and infinite observation periods for optimal demodulation. CPM sequences may be described by trellis representation, and use of the Viterbi algorithm reduces the observation period to a finite length. However, the complexity remains important and in practical applications, suboptimal solutions are normally implemented. The reduced complexity is traded-off against BER performance. In addition, the complexity increases with high M and L values and with low h-values, the same choices that would help us increase BER performance and spectral efficiency. CPM may be of use in some applications, especially if combined coding and modulation makes it possible to profit from the good CPM characteristics without being too complex [3].

In general, the memoryless modulation (PSK and QAM) would be more suited for packet transmission than modulation with memory (like CPM). The shorter the packets become, the more the truncation effect will degrade the reception.

3.1.2 Study of a Particular Modulation Scheme: Variable Rate N-MSK

The variable rate modulation scheme described, is based on a well-known continuous phase modulation scheme with a modulation index $h = 1/2$ and rectangular pulse shaping known as

MSK. If N-MSK signals, with different amplitudes are combined, the N-MSK signal is obtained. The N-MSK signal properties in linear and non-linear channels are described first; followed by analysis of the concept of switching among different modulation techniques. Finally some simulation results are presented.

3.1.2.1 Signal Description

The N-MSK signal is the sum up of n MSK signals with different amplitudes. The N-MSK can be described by following equations:

$$s_{N-MSK}\left(t, a_j^{(1)}, a_j^{(2)}, ..., a_j^{(n)}\right) = A \sum_{i=1}^{N} 2^{i-1} s_{MSK}^{(i)}\left(t, a_j^{(i)}\right) \tag{3.1}$$

where $s_{MSK}^{(i)}(t, a_i)$ is the ith MSK signal

$$s_{MSK}^{(i)}\left(t, a^{(j)}\right) = \sqrt{\frac{2E_b}{T}} \cos\left(2\pi f_c t + \pi a_j^{(i)} \frac{t}{2T} + \vartheta_j\right) \tag{3.2}$$

E_b is energy per bit, T is the duration of one bit, ϑ_j is accumulated phase in the jth time interval, f_c is carrier frequency and $a_j^{(i)}$ is the input data in the jth time interval and for i-th MSK signal. The N-MSK signal is normalised with the coefficient A. If $N = 1$ the 1-MSK signal is known as a MSK signal and if $N = 2$ the 2-MSK signal is named MAMSK signal. The scattering diagrams of N-MSK signal is plotted in Figure 3.4 for $N = 1$ and $N = 2$.

The signal spectrum is plotted in Figure 3.5. The same data rate is used for all calculated power spectra. The amount of signal bandwidth required for transmission of one bit is halved for 2-MSK signal in comparison with 1-MSK signal.

The MSK signal has a constant envelope, therefore non-linear amplification does not spread the signal spectrum. On the other hand, if $N > 1$ the envelope of signal is not constant, and spreading of the signal spectrum can be expected. However, the simulation results show that no significant signal spreading is observed if non-linear amplifier is used. An example of N-MSK signals amplified by TWT non-linear amplifier is plotted in Figure 3.6. The increase of a few dB is observed in the second, third and further power spectrum lobes.

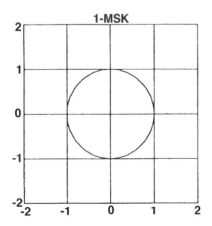
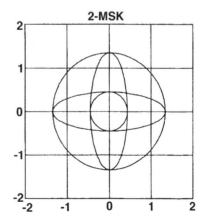

Figure 3.4 N-MSK signals scattering diagrams

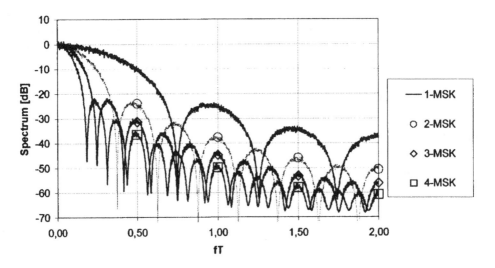

Figure 3.5   N-MSK signals power spectra, linear amplifier

3.1.2.2 Level Switching Approaches

Two level switching approaches are described in [12] for star QAM modulation scheme and one in [14] for variable rate CPFSK modulation schemes:

- Received signal strength indicator approach (RSSI);
- Error detector switching approach; and
- Eye closure switching approach.

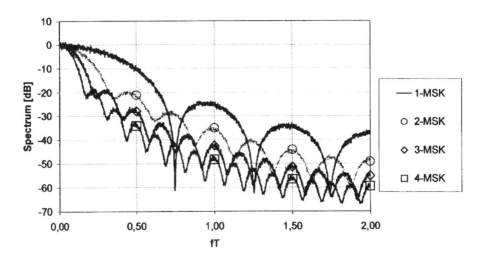

Figure 3.6   N-MSK signals power spectra, non-linear high power amplifier with operation point −3 dB

The first method is based on the received signal strength. The method assumes that the receiver contributes the majority of noise-to-signal before detection. If the signal is in deep fade, the received signal strength is small and the SNR ratio is low, therefore the BER is high for chosen modulation technique. Should the chosen service require better BER than that obtained by current received signal strength and chosen modulation scheme, e.g. 2-MSK, 2-MSK modulation schemes are switched to 1-MSK. However, if the specific data rate is also required, another burst in the same frame can be used for the same user. The choice of modulation scheme depends on receiver-generated noise – no other distortion is taken into account.

The second approach is the error detector switching approach, where the transmitted bits are encoded by error detection or error correction block codes. When an error is detected, the modulation scheme is changed to a more robust modulation scheme. If no error is detected in a certain amount of time, the modulation scheme is switched to a modulation scheme with higher bandwidth efficiency. The distortion of entire communication systems is mirrored in detected error, but the modulation scheme is changed after an error occurs.

The last approach is based on the observation of the eye closure at the detection points. Eye closure depends on distortions across the complete path of the transmitted signal and impairments of the transmission condition are detected before an error occurs.

The different frequency bands are used for up- and downlinks in the satellite systems and consequently the impairments in up- and downlinks are not correlated. For example, if a poor channel is detected in the downlink, it does not mean that the channel is poor in the uplink as well. The estimated channel characteristics should be transmitted from the receiver of the signal to the transmitter of the signal. The channel characteristics should not be significantly changed between two successive burst transmissions.

3.1.2.3 Simulation Model of Variable Rate Communication System

The simulation model of variable rate communication system is plotted in Figure 3.7. The burst structure is plotted in Table 3.1. The header of the burst consists of 32 bits, bearing the information necessary for the operation of a communication system. The robust MSK modulation scheme is used to transmit the first 32 bits in burst. The header is also used for received signal power estimation. Ninety-six symbols of the 128 are payload. The number of bits per symbol on the payload depends on channel quality.

The transmitted signal is distorted by Gaussian noise. The channel attenuation is variable and obeys log-normal distribution. The rate of channel variation is slower than the transmitted burst rate.

The eye closure approach is used as modulation switching criteria. The eye closure is defined as

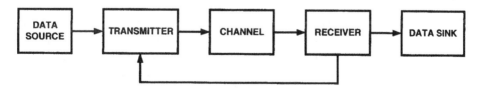

Figure 3.7   Simulation model of variable rate communication system

Table 3.1 Burst structure for variable rate communication system

Header	Payload
32 symbols	96 symbols
32 bits	96 or 192 bits

$$\text{eye}_{cl}[dB] = 10\log\left(\frac{1}{N}\sum_{i=0}^{N} s_T\ t_i) - s_R\ t_i))^2\right) \tag{3.3}$$

where $s_T(t_i)$ is the estimated signal at time t_i and $s_R(t_i)$ is the received signal at time t_i and N is the number of observed symbols. The eye closure is estimated at the moment when the eye opening is maximal. The BER dependence of eye closure is plotted in Figure 3.9.

Three different receivers are used in simulations: a conventional maximum likelihood receiver, a non-coherent receiver and a re-modulator receiver.

The Euclidean distance is calculated in one symbol interval for all possible transmitted signals in Maximum Likelihood Sequence Estimation (MLSE) receiver. The Viterbi detector is used to estimate the received signal when the difference from received signal is minimal. The received signal phase should be in phase with the signals generated at the receiver and the received signal must be correctly amplified for reliable results.

The high amplitude MSK component of the 2-MSK signal is determined by the signal phase in differential receiver. The low amplitude MSK component is determined by the signal envelope variation in successive symbol intervals. If significant amplitude variation occurs, data carried by low amplitude MSK component, will have an opposite sign to data, which is carried by high amplitude MSK component of the 2-MSK signal.

Data carried by a high amplitude MSK component of the 2-MSK signal is detected from the phase difference of the 2-MSK signal in successive symbol intervals. The estimated data is used to generate a high amplitude MSK signal. The re-modulated MSK signal is subtracted from received 2-MSK signal in the next step. The phase and amplitude of re-modulated MSK signals should correspond to the normalised received 2-MSK signal. If the data, carried by high amplitude MSK component of the 2-MSK signal, is not correctly detected, the increase in envelope of difference between 2-MSK and re-modulated MSK component is obvious. Increase in amplitude is used for the error detection and correction.

The Monte-Carlo method of simulation is used for finding BER. The minimum of 50 errors is sufficient for stopping simulation, if at least 10,000 symbols are transmitted.

3.1.24. Simulation Results

The 1-MSK and 2-MSK simulation schemes are individually tested in a Gaussian channel for different types of receiver. The BER curves are plotted in Figures 3.8 and 3.9. In Figure 3.8 the ratio between transmitted energy per bit and noise is used as an independent variable and the eye closer is plotted as abscissa in Figure 3.9. The performance of the 2-MSK signal is approximately 5 dB worse than the results for 1-MSK signal.

The results in Figure 3.9 are used to determine the switching level to shift between

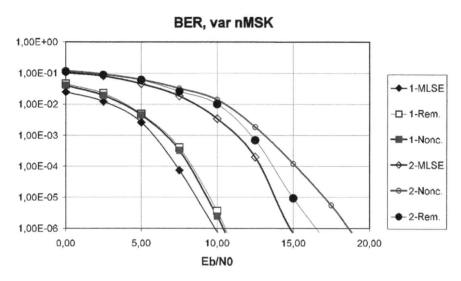

Figure 3.8   Bit error rate curves for N-MSK signals

modulation schemes. For example, if the BER better than 10^{-2} is required for a chosen communication service, the value of the eye closure corresponding to BER $= 10^{-2}$ can be estimated for 2-MSK signals. If the eye closure is less than -11 dB, the 2-MSK signal is transmitted in the next frame and if the eye closure is higher than -11 dB, the 1-MSK signal is transmitted.

The average number of transmitted bits and the BER rate for different receiver types are plotted in Figure 3.10. The variable σ denotes the variance of the channel attenuation. If σ is high, the probability of extreme poor channel is high and consequently the BER is increased.

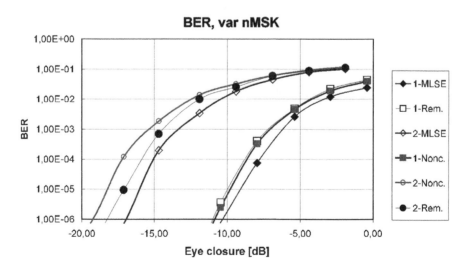

Figure 3.9   Eye closure influence on BER for N-MSK signals

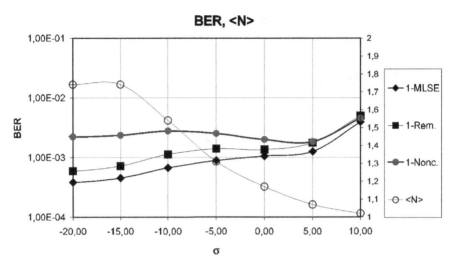

Figure 3.10   Average number of transmitted bits per symbol, if the receiver generated noise is 12.5 dB
lower than the signal average power

The average number of transmitted bits per symbol is close to one, because the probability of
low channel attenuation is low and the modulation is rarely switched to the 2-MSK modulation scheme. If the variance is increasing, the probability of high and low attenuation is
decreasing, and the average number of transmitted bits per symbol is increased and BER
is decreased. The results are plotted for receiver-generated noise 12.5 dB.

  If the noise, generated in the receiver, is increased to -10.0 dB the results for average
transmitted bits per symbol is decreased and BER is increased. The results are plotted in
Figure 3.11.

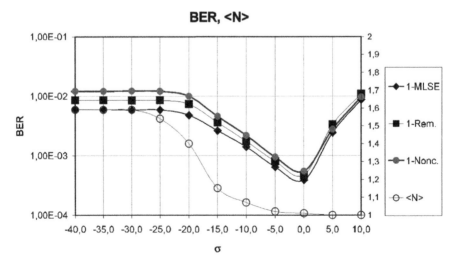

Figure 3.11   Average number of transmitted bits per symbol, if the receiver generated noise is 10.0 dB
lower than the signal average power

3.2 Coding Techniques

The Shannon theorem states that every transmission channel has a capacity C (in bits/s) such that for any information transfer rate $R < C$, there exists a code with length n giving a bit error probability $P < 2^{-n E_b^{(R)}}$. $E_b(R)$ is a positive function of R and determined solely by the channel characteristics. The demonstration is based on the random coding argument. For a bandwidth-limited AWGN channel we have:

$$C/W = \log_2(1 + (E_b/N_o)(C/W)) \tag{3.4}$$

where W is the bandwidth, E_b the binary energy and N_0 the noise power spectral density.

Most channel coding techniques are constructed with the same aim: to reduce the BER in an AWGN channel with tolerable decoding complexity. The codes are constructed for AWGN channels and are usually useless when confronted with non-linearities, burst errors and fading environment. Other techniques are implemented in order to take care of those impairments, such as predistortion techniques for non-linearities, interleavers for burst errors and equalisers for fading channels, with the aim to whiten the noise and distribute the errors before decoding. In many cases, such as mobile communication, acknowledgement protocols, retransmission and channel information also increase the quality of the transmission at the expense of reduced system efficiency. Some of the most common coding schemes are presented, without considering any return channel for acknowledgements or side information.

The important characteristics of the codes are considered to be their fundamental coding gain – comprising the coding gain and the spectral efficiency, and the complexity. The fundamental coding gain was introduced by Forney in [4] for coset codes, but is easily extended to the following definition:

> The fundamental coding gain is the coding gain between the SNR necessary for the achievement of a specific BER with channel coding, and the SNR necessary for obtaining the same BER with an unmodulated scheme carrying the same number of bits per symbol period.

The unmodulated scheme will normally be a fictive modulation since the number of bits/T_s of a coded scheme is non-integer in general, however the interpolation of unmodulated schemes makes such a comparison possible, and the fundamental coding gain definition has the advantage of including spectral efficiency when comparing codes. The extension of the fundamental coding gain definition to codes other than coset codes, makes the gain measure more approximate, since bit mapping and other implementation issues can influence the coding performance. The definition also gives a helpful indication on the performance gain.

3.2.1 Overview of Channel Codes for Satellite Transmission

As stated in the introduction, the traffic type used in the system has a direct impact on the choice of code. Codes requiring decoders which need long observation periods for optimal decoding (e.g. convolutional codes or concatenated codes) demonstrates important performance degradation depending on packet lengths and code parameter choices. This section provides an overview of the most common channel code families, and discusses their suitability for our system.

3.2.1.1 Linear Block Codes

Linear block codes are usually thought of as BCH (Bose–Chaudhuri–Hocquenghem), RS (Reed–Solomon) or Hamming codes. Actually, both RS and Hamming codes are part of the general BCH code family [5]. In addition to these codes, there are many other linear block codes, which will not be covered. The BCH code theory is based on linear algebra in the Galois Field GF(2^q), and the goal of these codes is to increase the minimum distance, d_{min}, between q-ary code words. The separation of the modulation and the coding implies a non-optimal mapping between code sequences and modulated symbol sequences. The decoding is usually performed by hard decision for complexity reasons. The fundamental code gain obtained by BCH codes ranges from fractions of dB to a few dB. The coding operation adds redundancy which is sent over the channel with the information, thus increasing the required bandwidth of the signal. The reduction in spectral efficiency is k/n, where k and n are the number of information and coded symbols, respectively. The correction capacity is $t \leq (d_{min} - 1)/2$. The general BCH code word is illustrated on Figure 3.12.

In packet switched satellite communication, the block format of the linear block codes fits well to the packet traffic type. However, the coding gains obtained with linear block codes fail to meet the requirements of the future (and even of today's) satellite systems – high quality at low SNR, bandwidth efficiency and low complexity.

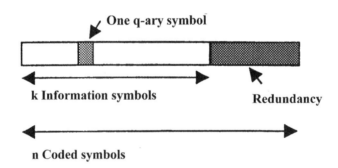

Figure 3.12   Usual organisation of a linear block code based on GF(2^q). If the code is binary $q = 2$

3.2.1.2 Convolutional Codes

Contrary to block codes, convolutional codes are semi-infinite codewords; otherwise, they share many of the properties of block codes – they are binary codes aiming at increasing the minimal distance between sequences. The need of a mapper between encoder and modulator usually results in hard decisions at the decoder, unless the number of simultaneously coded bits at the output of the encoder corresponds to one point in the signal constellation. The code sequence is generated by a shift register with binary operations on the contents according to the generator polynomials [5]. The code sequence may be described by a state trellis, and hence, the decoding is easily performed by the Viterbi algorithm. The Viterbi algorithm is a MLSE with quasioptimal performance even for truncated observation periods. Also in this case the spectral efficiency is reduced by k/n, where k and n are the number of input and output bits of the shift register, respectively. The coding gains obtained with convolutional codes are usually of the same order of magnitude as those obtained with linear block codes.

The structure of the convolutional code does not suit the packet format, even when the packets are small. In this situation, use of convolutional codes will result in suboptimal performance, either because of truncation of the decoder algorithm, need of code state information or flushing for trellis termination. When a high gain is required, the number of trellis states must also be high, and the decoder for the same packet length will be suboptimal.

Figure 3.13 shows the trellis representation of a four-state convolutional code.

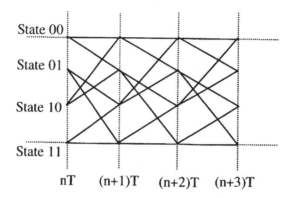

Figure 3.13   Trellis diagram of a rate 1/2 four-state convolutional code. The input bit determines the branch taken from a particular state at a certain moment. The solid lines represent the possible code-words of this code. Each branch represents two output bits

3.2.13. Concatenated Codes

One popular means of obtaining high coding gains with relatively low complexity, is to employ concatenated coding. Convolutional codes, together with RS codes, are often suggested (see Figure 3.14). Coding gains of several dB can be achieved, and these schemes are also proven to be efficient for satellite communication [6]. However, the maximum code gains are obtained only if the error patterns at the input of the decoders correspond to the isolated, evenly distributed errors in an AWGN channel. This is normally not the case at the decoder output, and therefore interleaving is usually required between the two codes in order

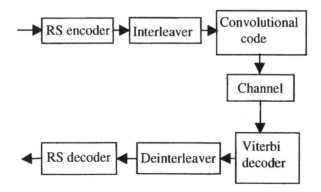

Figure 3.14   Functional representation of concatenated coding with deinterleaving for spreading of error bursts present at the Viterbi decoder output

to disperse the error bursts. The use of an interleaver requires interleaving of several code-words of the outer code in order to spread the error bursts on a sufficiently large number of codewords. If the code word length of the outer code corresponds to the packet size, several packets must be interleaved. Multiplexing packets from several sources will then become a problem, and severe reductions in code performance will result. This is why concatenated codes are not considered the best solution for packet switched satellite communication.

3.2.1.4 Coded Modulation

Coded modulation was introduced in the 1980s [7], and opened a new area in coding theory. The concept of combining modulation and coding in order to optimise the Euclidean distance of the code words instead of the Hamming distance, made trellis coded modulation very promising, both from a theoretical and practical point of view. Soft decision decoding became inherently optimal, and coding gains on the order of 3–6 dB were achievable. The decoder complexity increased compared with traditional codes due to the soft decision, otherwise the decoding is based on the trellis algorithm as for convolutional codes. Research was carried out for coding, and many proposals for Trellis Coded Modulation (TCM), and Block Coded Modulation (BCM) were published. The main idea behind this code concept is to increase the Euclidean distance between coded sequences by two methods. The first is to increase the distance by subset partitioning (see Figure 3.15), which means that a signal constellation is divided into two subsets at every partitioning level, and there are m levels. The distance between the constellation points increases by a factor of $\sqrt{2}$ at each partition level if we take M-QAM as an example. When the target distance has been achieved, the m bits representing a certain partition will be the encoder output bits, and the subset points will be represented by the uncoded bits. Therefore, the code optimisation consists of choosing a code in such a way as to select the subsets in the right order so that the distance between sequences in the trellis will not be much lower than the distance between points within a partition subset. If the chosen encoder has m output bits for $m - 1$ input bits, the coded modulation will employ a

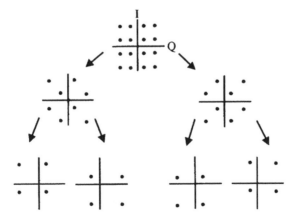

Figure 3.15 Two-level set partitioning of a 16-QAM. Two coded output bits of a 1/2 convolutional encoder designate one of the subsets, while two uncoded bits designate one of the constellation points within the subset. The spectral efficiency is 3 bits/T_s

constellation containing double the amount of constellation points than the uncoded modulation carrying the same number of useful bits per symbol (i.e. a 32-QAM will carry 4 bits/T_s, the same as an uncoded 16-QAM). This means that there will be no bandwidth expansion related to the use of a coded modulation. The price to pay is higher signal power for a higher order modulation scheme, a power which is usually more than recovered by the coding gain. The 3–6 dB coding gain referred to is the fundamental coding gain, which is comparable to a real modulation scheme in this case (at least for two-dimensional coded modulation).

Coded modulation are good codes, and worth considering for satellite communication. The BCM would perhaps be the most appropriate for packet transmission. Unfortunately, the coding gains remain a bit too low considering what is normally required in satellite communication.

3.2.1.5 Turbo Codes

Turbo codes are the latest newcomer within coding [8], and are considered as a kind of revolution within the channel coding community. These codes attack the problem of approaching the Shannon limit from a different angle than for coded modulation. This time the randomness and the code length aspects in the Shannon theorem are met. Together with an iterative soft decision algorithm, this asymptotically brings the code performance very close to the Shannon limit – in some cases about up to 10dB coding gain. The first Turbo codes proposed were based on Parallel Concatenated Convolutional codes (PCC). Two or more convolutional codes are implemented in parallel, the input bitstream is the same for each encoder, only permuted by an interleaver in front all except the first (see Figure 3.16). The output is a multiplex of the encoder outputs, which may or may not be punctured. The decoder will operate on the incoming stream with soft decisions for the first decoder, and the soft output of this decoder is used as side information and help for the second decoder (after

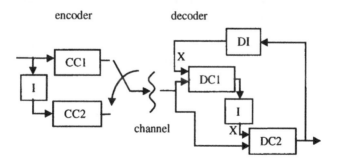

CC: Convolutional encoder
DC: Decoder
I: Interleaver
DI: Deinterleaver

X: Extrinsic information

Figure 3.16   Typical example of a Turbo encoder/decoder

deinterleaving), etc. When all the decoders have made their decision, it is possible to feed back this information to the first decoder, and start all over again, only that now the side information is much more reliable than in the first decoder iteration. It is possible to continue like this indefinitely, but for each iteration, the extra gain obtained is less and less significant. The practical limit is when time delay and/or coding gain is estimated sufficient.

Since the pioneering work on PCC, several other types of Turbo codes have been proposed, and it has been recognised that the soft input–soft output iterative decoding principle may be employed with success on several types of codes. We can mention proposals in the direction of serial concatenated codes (SCC), structured instead of pseudo-random interleavers (in order to shorten the code word lengths) and Turbo decoding of Product Codes (TPC). We have investigated the latter for use in satellite packet switched transmission for the following reasons:

- The TPC codes are well adapted to block codes;
- There is an inherent interleaving in the TPC structure;
- Packet sizes may be small;
- The coders and decoders involved are simple BCH codes.

In the following section, the TPC technique, adaptation to traffic types of interest, and simulation results on their performance are presented.

3.2.2 Study of a Particular Channel Code – TPC

We want to propose 'Turbo decoded product codes' as described by Pyndiah in [9]. The most convenient modulation scheme for this code is the well-established QPSK modulation. The reason is that for this scheme, the binary mapping and the consequent Hamming distances may to some extent be reflected in the Euclidean distances if Gray mapping is used. This is strictly true for a BPSK modulation, but BPSK is inefficient for this type of satellite communication. The Euclidean distance concept is important for soft decoding on which the Turbo decoding is based.

Coders and decoders are based on simple one- or two-error correcting BCH codes. Their good performance is due to the inherent interleaving of product codes, the soft input/soft output and the iterative decoding principle of the Turbo codes. The non-real-time character of some satellite links means that this type of traffic should not suffer from delay impairments due to the iterative decoding. The delay imposed by the decoder is on the order of 3–4 packets. If the bit rate is below ~10 Mbit/s, accelerated processing is possible at the receiver, thus enabling four matrix decoder iterations in one packet period.

A short review of the principles of this code, resumed from [9] will be presented, followed by some of our simulation results.

3.2.2.1 Code Principle

The idea of this coding scheme is to use the well-known product codes, which may be put in a matrix form for two-dimensional coding, or in a cubical form for three dimensions etc. We will concentrate on two dimensions in the following, but the implementation presented in [10] integrates 3 and 4 dimensional codes also. The matrix form of the two-dimensional code is depicted in Figure 3.17. The k_1 information bits in the rows are encoded into n_1 bits, by using

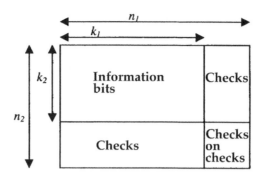

Figure 3.17   Two-dimensional product code matrix

i.e. a $BCH_1(n_1, k_1, t_1)$ code, where t_1 is the correction capacity of the code, $r_1 = n_1 - k_1$ the redundancy and d_1 the Hamming distance. After having encoded the rows, we proceed to coding of the columns using a $BCH_2(n_2, k_2, t_2)$, where also the check bits of BCH_1 are encoded. A more general denomination is to call BCH_1, C_1, and BCH_2, C_2, since the codes may be any binary block code. We have chosen to use the BCH example in this presentation. It is clear that the global number of information bits of this product code is $k_1 \times k_2$, the global code rate is $R = R_1 \times R_2$, where $R_i = k_i/n_i$. Another property is that the global Hamming distance is $d = d_1 \times d_2$.

The decoding procedure is as follows:

- We may assume that we start by decoding the matrix rows. Soft input is calculated by squared Euclidean distance calculation on the I and Q components of the QPSK signal. A certain radius p around each received codeword is chosen as a parameter. The p least reliable positions in the received word are determined, and by a set of test-sequences formed by the p least reliable positions, the codewords lying closest to the received word are determined by hard decision BCH decoding. This algorithm is called the Chase algorithm.
- A weighted reliability on each of the bits in the received word is calculated by simplified derivatives of the LLR (Log-Likelihood Ratio) algorithm. This weighted reliability on each decision is used as the soft output of the row decoding, which is identical to the soft input for the decoding of the columns. The same algorithms are used for the columns, and we may return to the rows for a second decoding, etc.

If the product code is chosen so that BCH_1 is the same as BCH_2, we have a square product code, which simplifies the decoder since only one BCH decoder is needed.

3.2.2.2 Performance

The main idea for this code proposition is to avoid problems connected with suboptimal design of concatenation due to interleaving requirements and it is clear that the Turbo Product Codes must be designed in order for one matrix to fit one packet length. A common packet length is the 53-byte ATM cell or a multiple of ATM cells. Table 3.2 gives an indication on adequate code candidates if we consider packet lengths of one and two ATM cells. N_b is the

Table 3.2   Squared product code adaptation to packet lengths

Packet	N_b	N_b	BCH $(n,k,t)^2$	BW e.f.	BW e.f. shortened
1 ATM cell	424	20.6	$(31,26,1)^2$	1.42	1.67
			$(31,21,2)^2$	2.18	2.23
2 ATM cells	848	29.1	$(63,57,1)^2$	1.22	1.85
			$(63,51,2)^2$	1.53	2.61

number of bits in one packet, $\sqrt{N_b}$ the number of bits in each dimension in an ideal square matrix.

As we can see from Table 3.2, the square root number of bits in the packets does not fit into a simple BCH code scheme correcting 1 or 2 errors. It is however easy to shorten the BCH codes, and we have proposed some simple BCH codes in Table 3.2 that may be applied to the packets. It should be noted that it is not the Hamming distance of the constituent codes that matters but the distance of the product code. It is possible to design codes with almost any error correcting property by selecting different constituent codes. Squared codes here for simplicity have been chosen. In Table 3.2 we have also reported the bandwidth expansion factor of the product codes (BW e.f.), and since these codes should be shortened, we have also given the according bandwidth expansion factor (BW e.f. sh.). For example, the $(31,26,1)^2$ code must be shortened by $b = 26^2 - 424 = 252$ bits in order to fit one ATM cell, giving a bandwidth expansion factor equal to $(n - b)/(k - b) = 1.67$ instead of $n/k = 1.42$. The performance results given in Figure 3.18 correspond to the unshortened codes, and the curves must be corrected with a shift towards higher E_b/N_0 values due to the shortening. In the previous example, this shift equals $10\log(1.67/1.42) = 0.7$ dB. All the results shown correspond to 4 matrix decoding iterations.

The results given in the literature and by our simulations show two important properties:

- Good performance is obtained with simple squared codes correcting two binary errors.
- Good performance is obtained after only 4 matrix decoding iterations.

3.3 Synchronisation

Synchronisation is a huge area because it is involved at many levels. The carrier frequency, the symbol clock rate, the frame rate, the phase ambiguity, and the code rate, must be synchronised. These characteristics are often completely or partly unknown to the receiver. How to proceed with the synchronisation depends on the traffic type and application. The more information lost for synchronisation purposes, the better the synchronisation will be, both in terms of accuracy and maintenance. However, the transmission is packet based, and short messages require immediate synchronisation without packet loss.

3.3.1 Overview of Required Synchronisation for Satellite Transmission

The immediate synchronisation without packet loss is a big challenge in satellite systems,

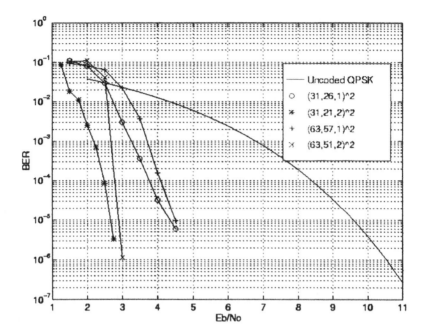

Figure 3.18   BER vs. E_b/N_0 in dB for uncoded QPSK and Turbo-decoded square product codes based on BCH codes

characterised by a noisy channel with non-linearities. Much research has been carried out on synchronisation for satellite transmission. New challenges will be encountered as the SNR decreases due to capacity claims, and enhanced channel coding techniques.

However, typical synchronisation problems are not covered in this report – other aspects of synchronisation, unique to mobile systems have been studied. In our case, the satellites are mobile, and although the terminals are also mobile, the motion of the satellite is responsible for the majority of mobility. The mobile systems are prone to Döppler frequency shifts, which are deterministic, but when the problem is to determine how and where the correction should be made including aspects of complexity; the correction should be in the transmitter, the receiver or the satellite. The choice not only has an impact on the complexity of the equipment, but will also influence the network management through determination of spot-beam sizes, frequency reuse etc. In this part of the report, we have concentrated on how to perform a Döppler frequency correction in the receiver. Three techniques have been studied, the classical feedback loop for reference, and two feedforward loops. The techniques are evaluated in terms of performance, acquisition time and range, and jitter. The entire study is based on normalised values, and is therefore general. It is clear that the Döppler shift is much more important relative to low symbol rates that to high symbol rates.

3.3.2 Study of Döppler Frequency Shift Compensation for Mobile Satellites

In this section, the context of our study and the choice of algorithms for analysis will be presented, followed by a description of the main characteristics of our modelling.

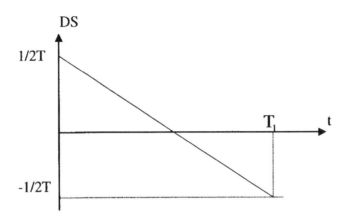

Figure 3.19   Döppler shift variation as a function of time

The Döppler shift may vary continuously from a positive value to the equivalent negative value with time. The shift is be expressed in terms of fractions of the symbol rate, as illustrated on Figure 3.19. We have chosen the value $1/2T$ as a reference value for a typical Döppler shift. This would correspond to the values encountered in a system using LEO satellites moving at a speed of 7.5 km/s and a symbol rate transmission of about 100 kbauds. T_1 corresponds to the period during which the terminal is connected to a satellite. After T_1, the communication moves to a new ascending satellite for another period T_1. In our numerical example $T_1 = 4 \times 10^7$ symbol periods. The terminal will experience an abrupt variation of the Döppler shift from the maximum negative value $-1/2T$, to the maximum positive value $1/2T$.

The satellite beam may be divided into spotbeams in order to reduce the Döppler shift which must be corrected by the terminal. This division requires spotbeam handovers. If the satellite beam has been divided into p spotbeams, the carrier frequency of the kth beam is shifted by $f_k = 1/2T + (3 - 2k)/2pT$, with k ranging from 0 to $p - 1$, and the carrier frequency will vary from $f_0 + f_k + 1/2pT$ to $f_0 + f_k - 1/2pT$ within each beam in a T_1/p period. The division by p of the extreme values of the Döppler shift permits the use of lower bitrates, or a quicker re-acquisition.

The channel degradation is illustrated by an AWGN signal and a DS $\Delta\Omega$. $h(t)$ and $g(t)$ represent the transmitter and receiver filter, respectively, forming a Nyquist filtering globally, and initially equally distributed between the two. The influence of the Döppler shift on the signal, is demonstrated by the complex multiplication with $e^{j\Delta\Omega t}$, and due to this shift, the global Nyquist filter no longer guarantees ISI-free transmission. The correction of one channel impairment is not always possible without an enhanced degradation due to another. The aim of this study is to identify the effects of the different algorithmic parameter choices and to optimise the signal reception.

The expression of the base-band transmit signal is written:

$$x(t) = \exp\{ - j\Phi(t)\} \tag{3.5}$$

where $\Phi(t)$ is the phase carrying the PSK information. The signal at the output of squared Nyquist filter is written:

$$y(t) = h(t) * x(t) \tag{3.6}$$

where $*$ denotes convolution. After transition through the channel, the signal becomes:

$$z(t) = y(t)\exp\{j\Delta\omega t\} + n(t) \tag{3.7}$$

where $n(t)$ denotes the noise with a Spectral Power Density (SPD) equal to N_0 and $\Delta\Omega$ is the Döppler frequency shift.

We have chosen to study three compensation techniques. First, as a reference, we will study a classical feedback loop method based on phase comparison between the received symbol and the decided symbol. The second technique is an open loop method based on differential detection [15]. This algorithm will be shown to be particularly robust to large Döppler shifts. Finally, we will study the performance of the double differential detection [17], performing the quickest re-acquisition.

3.3.21. Algorithmic Description

The classical feedback algorithm chosen as a reference is based on the comparison between the received signal phase and the phase of the decided QPSK symbol. At the input of the receiver, after the square Nyquist filter, the signal $z(t)$ becomes:

$$u(t) = g(t) * z(t) \tag{3.8}$$

In an ideal case (noise- and ISI-free):

$$u(t) = \exp\{j(\Delta\omega t + \Phi(t))\} \tag{3.9}$$

where $\Phi(t)$ is the phase shift due to QPSK modulation. Then $u(t)$ is corrected by a signal coming from a voltage controlled oscillator (VCO) and becomes:

$$v(t) = \exp\{j(\Delta\omega t + \Phi(t))\} \times \exp\{-j\Delta\omega t\} \tag{3.10}$$

$$= \exp\{j(\Delta^2\omega t + \Phi(t))\} \tag{3.11}$$

where $\Delta\omega' t$ is an estimation of $\Delta\omega t$ and

$$\Delta^2\omega t = (\Delta\omega t - \Delta\omega' t) \tag{3.12}$$

Let $s(t) = \exp\{j\check{\Phi}(t)\}$ the QPSK symbol after hard decision, the phase detector output is written:

$$\varepsilon(t) = \mathrm{imag}(v(t) \times s * (t)) = \sin(\Delta^2\omega t + \Phi(t) - \Psi(t)) \tag{3.13}$$

where imag denotes the imaginary part of a complex, and $*$ denotes conjugation. Suppose that $|\Delta^2\omega t|$ is smaller than $\pi/4$, and without noise, $\Psi(t) = \Phi(t)$. After a phase-frequency detector, $\varepsilon(t)$ is low-pass filtered and the control signal at the Voltage Controlled Oscillation (VCO) filter is $\exp\{-jd(t)\}$ giving the estimation:

$$\Delta\omega' t = d(t) \tag{3.14}$$

We have chosen a Nyquist filter with $\alpha = 0.3$ for our simulations. The Döppler shift introduced in the channel will cause degradations due to filter mismatch between the transmitter and the receiver. This mismatch may be avoided by introducing the filter after the frequency correction. This solution may also be delicate due to the delay introduced in the loop which causes instability. This topic is not the object of our study and we have chosen the first solution with the Nyquist filter outside the loop.

The first differential algorithm studied was an open-loop technique for estimating and correcting Döppler frequency shift in a Quadrature Differential Phase Shift-Keyed (QDPSK) receiver. This technique, based on the fact that the change in phase over half a symbol contains only the Döppler induced phase shift, has been covered by Simon and Divsalar in [15].

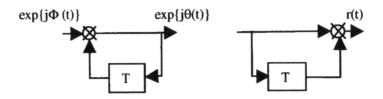

Figure 3.20   Differential detection principle

The conceptual differential scheme is illustrated in Figure 3.20, where $\exp\{j\Phi(t)\}$ is the QPSK symbol, and $\exp\{j\theta(t)\}$ the QDPSK symbol. $r(t)$ is the differentially detected signal at the receiver. Differential detection is more power-efficient and robust than coherent detection in digital communications applications where carrier phase and frequency are uncertain. Phase tracking loop is unnecessary and difficulties of acquisition and maintenance of phase and frequency lock are avoided. The differential encoder induces a controlled inter-symbol interference, which will be displayed at the receiver for demodulation. The received signal is in this case:

$$z(t) = y(t)\exp\{j\Delta\omega t\} + n(t) \tag{3.15}$$

with:

$$y(t) = h(t) * [x(t) \times x(t - T)] \tag{3.16}$$

In [15], it is shown that ISI-free filtering can be obtained by using a 100% excess bandwidth squared Nyquist filter (roll-off factor $\alpha = 1$), with a filter described as a pseudo-matched filter at the receiver. This pseudo-matched filter consists of a brickwall (BW) filter and the addition of two $T/2$-spaced samples. These samples are selected at a particular moment in order to obtain the ISI-free equivalence. With this filter, the ISI-free transmission is acquired with a noise penalty of 3 dB due to the 100% excess bandwidth.

The algorithm proposed in [15] is based on the fact that the change in phase over half a symbol contains only the Döppler shift information if the pulse shape $\theta(t)$ was chosen such that $\theta((k + 3/4)T) = \theta((k + 1/4)T)$. The pulse-shape condition is obtained by the squared-root Nyquist transmit filter with the pseudo-matched filter at the receiver. The estimation of the Döppler frequency shift is obtained by multiplying two $T/2$-spaced samples. Every multi-plication and addition leads to additional noise, which is limited if not eliminated, in the Döppler estimation by a LowPass Filter (LPF). If the Döppler shift is constant in time, the ideal input (without noise) to the LPF is a constant signal (Figure 3.21).

Without noise, the Döppler estimator processes as follows:

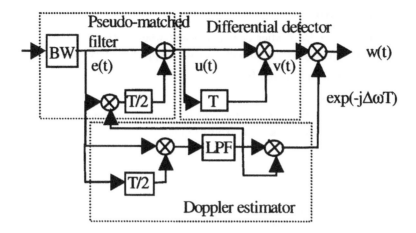

Figure 3.21   Döppler shift correcting receiver

$$e((k + 3/4)T \times e * ((k + 1/4)T) = \exp\{j(\Delta\omega T/2 + \theta((k + 3/4)T) - \theta((k + 1/4)T))\}$$

$$= \exp\{j\Delta\omega T/2\} \tag{3.17}$$

If $e(t)$ denotes the baseband signal at the BW filter output, and $*$ the conjugation. This estimation of the Döppler shift can be used to correct the exit of the differential detector. After normalisation, multiplication by itself and conjugation, the signal becomes $\exp\{-j\Delta\omega T\}$. This estimation of the Döppler shift can be used to correct the exit of the differential detector.

The signal at the receiver input is $z(t)$ defined previously. After the BW filter, it becomes:

$$e(t) = g(t) * x(t) \tag{3.18}$$

sampled at $t = (2n - 1)T/4$ and where $g(t)$ is the temporal expression of the BW filter:

$$g(t) = \text{sinc}(\pi Bt) \tag{3.19}$$

where B is the chosen bandwidth. If we considered the ideal case (noise free and negligible distortion of the brick-wall filter):

$$e(t) = \exp\{j(\Delta\omega T + \theta(t))\} \tag{3.20}$$

where $\theta(t)$ is the phase shift due to QDPSK modulation. Thus, after pseudo-matched filtering, the signal at the differential detector input is:

$$u(t) = e(t) + e(t - T/2) \times \exp(j\Delta\omega T/2)$$

$$= 2 \times \exp\{j(\Delta\omega t + \theta(t))\} \tag{3.21}$$

with samples at $t = (k + 3/4)T$ after decimation (with one sample per symbol period) and where $\exp(j\Delta\Omega T/2)$ comes from the Döppler estimator. This signal, $u(t)$, is ISI-free and contains the Döppler shifted DPSK symbols.

The expression of the differential detector output signal is (with $u(t)$ normalised):

$$v(t) = u(t) \times u * (t - T)$$

$$= \exp\{j(\Delta\omega T + \Phi(t))\} \tag{3.22}$$

with samples at $t = (k + 3/4)T$ after decimation and where $\Phi(t) = \theta(t) - \theta(t - T)$ is the QPSK symbol phase. Finally, $v(t)$ is corrected by the Döppler shift estimation and becomes:

$$w(t) = v(t) \times \exp\{-j\Delta\omega T\} = \exp\{j\Phi(t)\} \tag{3.23}$$

which is the QPSK symbol.

The BW filter is a perfect bandpass filter, which is centred on the carrier frequency. In a baseband notation, its frequency domain expression is:

$$G(f) = 1 \qquad \text{if } |f| < B/2 \tag{3.24}$$

$$G(f) = 0 \qquad \text{otherwise.}$$

The size of this filter bandwidth is at least the size of the transmitted signal: $B = 2/T$. When a Döppler shift is present, two alternatives are possible: the filter bandwidth is adapted to the transmit filter and suffers from degradation due to filter mismatch; or widen B to include the Döppler shift and thereby increase the noise level.

The noise sensitivity of the Döppler correction branch is relatively high, which implies that the bandwidth of the LPF must be as small as possible. As previously indicated, the Döppler correction signal is a constant if the Döppler shift is constant in time. In this case, the LPF bandwidth may be chosen with changing small bandwidth in order to filter the noise to as high an extent as possible. However, when the Döppler frequency shift is varies, the choice of the filter bandwidth of the LPF must be considered. The larger the bandwidth, the higher the acquisition range of the Döppler shift, but the more the receiver performance will suffer from noise degradation in the Döppler estimator. A compromise will depend on an appropriate choice of the parameter χ of the filter chosen in our study (from [16]). The transfer function of this filter is:

$$L(\omega) = (1 - \chi)/(1 - \chi e^{-j\omega T}) \tag{3.25}$$

Thus, the filter gain is $Q(0) = 1/(1 + \chi)$ and the filter half bandwidth f_b should verify the following equation:

$$Q(2\pi f_b) = 1/(2(1 - \chi)) \tag{3.26}$$

giving with first order approximations for $\chi \to 1$;

$$f_b \approx (1 - \chi)F_s/(2\pi) \tag{3.27}$$

where F_s is the data rate.

The second open loop technique studied for estimating and correcting Döppler shift, is based on second-order phase difference modulation [17].

The conceptual double differential scheme is illustrated in Figure 3.22 where $a(t) = \exp\{j\Phi(t)\}$ is the QPSK symbol, $b(t) = \exp\{j\vartheta(t)\}$ the DPSK symbol and $c(t) = \exp\{j\theta(t)\}$ the DDPSK symbol.

We will illustrate that the double differential scheme allows differential detection, which is more efficient and robust than coherent detection, even if the frequency is shifted by Döppler. There is however, a high noise correlation due to the share of a noise sample at $(t - T)$ [16], a 3–4 dB degradation in comparison with coherent QPSK according to [17]. Modifying the

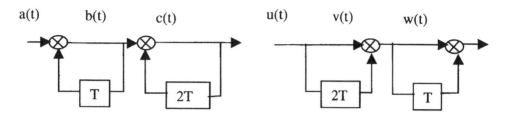

Figure 3.22 Principle of DDPSK transmitter and receiver

algorithm can significantly reduce this degradation. The only modification needed, is to change the second delay element at the transmitter (and correspondingly the first at the receiver) into a $2T$ delay element. It can be shown, by calculating the noise SPD, that there are no longer any noise terms due to cross-correlation. Simulations confirm the results of the previous analysis that a significant portion of the performance degradation can be avoided with this small algorithmic change and with only a small delay penalty.

The expression of the complex baseband transmit signal (before the differential modulation) is written as:

$$a(t) = \exp\{j\Phi(t)\} \tag{3.28}$$

where Φ is the phase carrying the PSK information. The baseband transmitted signal, after double differential modulation and Nyquist filtering may be written:

$$y(t) = h(t) * c(t) = h(t) * [x(t) \times x(t - T) \times x(t - 2T)] \tag{3.29}$$

where $h(t)$ is the Nyquist filter. Thus, after transit through the AWGN channel, the base-band signal at the receiver input is:

$$z(t) = y(t)\exp\{j\Delta\omega t\} + n(t) \tag{3.30}$$

where $n(t)$ denotes the complex noise signal with a SPD equal to N_0, and $\exp\{j\Delta\Omega t\}$ represents the Döppler shift of the signal. The signal at the receiver input, $z(t)$, become after the BW filter:

$$u(t) = g(t) * x(t) \tag{3.31}$$

where $g(t)$ is the BW filter.

If we consider the ideal case (noise free and without any distortion caused by the BW filter):

$$u(t) = \exp\{j(\Delta\omega t + \theta(t))\} \tag{3.32}$$

where $\theta(t)$ is the second order phase shift due to QDDPSK modulation.

The expression of the signal at the first stage of the double differential detector is (considering only the phase of $u(t)$):

$$v(t) = u(t) \times u * (t - 2T)$$

$$= \exp\{j(2\Delta\omega T + \vartheta(t))\} \tag{3.33}$$

where $\vartheta(t) = \theta(t) - \theta(t - 2T)$ is the QDPSK $(T,2T)$ symbol phase. The expression of the signal at the output of the double differential detector is (considering only the phase of $v(t)$):

$$w(t) = v(t).v * (t - T)$$

$$= \exp\{j\Phi(t)\} \tag{3.34}$$

where $\Phi(t) = \vartheta(t) - \vartheta(t - T)$ is the phase of the QPSK information-carrying symbol.

Thus, we see that a double differential algorithm is based on the fact that the Döppler shift term, which becomes constant after one differential detection, disappears after the second differential detection.

The choice of the roll-off factor α of the Nyquist transmit filter will characterise the shape of the pulse which is important when the carrier frequency is shifted by Döppler. If the BW filter is designed to match the transmitted signal spectrum exactly, with as low a penalty to noise degradation as possible, a small value for α would be suitable. In the presence of a Döppler shift, the consequence would be an important degradation due to the power loss induced by the filter mismatch. In order to avoid this, a higher value of α may be chosen, but this time, the noise bandwidth penalty would increase with increasing values of α. If α is close to 1, the BW filter bandwidth will have to be adapted to the spectral size of the transmitted signal, and the integrated noise level will be vital.

3.3.2.2 Performance Analysis

The performance analysis will be carried out according to three criteria. First, we will study degradation in terms of bit error rate, which will be sensitive to Döppler shift and noise. Secondly, the acquisition range and time will be studied when the Döppler shift varies. Finally the root mean square of the Döppler estimator, and its sensitivity to timing jitter, will be evaluated since filter mismatch cause ISI, and the optimal sampling instant may change.

Classical Feedback Loop BER Performance

Simulation shows that a constant Döppler shift of $1/20T$ does not cause any noticeable degradation to the BER; whereas a constant Döppler shift of $1/10T$ causes a degradation of 0.5 dB at a BER of 10^{-3}. It also shows that for DS $= 1/10T$ and a low signal-to-noise ratio (lower than 6.5 dB), 10,000 symbols are insufficient for the system to acquire loop lock. Thus, we already see that this type of frequency detector will not be suitable for systems needing fast acquisition.

Classical Feedback Loop Acquisition Range and Time

Simulations show that a Döppler shift of about $1/10T$ is the maximum which can be tolerated for low SNR (about 7dB). The upper limit, $1/8T$, is found by simulation without noise. Our simulations also indicate that the acquisition time is about 2000 symbols if the Döppler shift is $1/20T$ and about 10,000 symbols if DS $= 1/10T$. This acquisition time seems quite independent of the noise level.

In our system, the Döppler shift varies from $1/2T$ to $-1/2T$ on $T_1 = 4 \times 10^7$ symbol periods. It is therefore necessary to divide the beam by at least five or ten in order to achieve acquisition. First, we will indicate how the oscillator tracks the Döppler shift when it varies linearly and, secondly, we will discuss the effect of hand-over, in terms of acquisition time, when one beam is changed into the next. We consider that the Döppler shift varies linearly from $1/10T$ to $-1/10T$ in 8.10^6T, or from $1/20T$ to $-1/20T$ in 4.10^6T. Simulations also show that this linear variation of the Döppler shift does not cause any additional degradation on the BER, when compared with the constant Döppler shift case.

As mentioned previously, acquisition is only possible if $\Delta\Omega T < \pi/4$ (due to QPSK signal shifting constellation quadrant with higher phase values). If we consider a jump from $-1/10T$ to $1/10T$, the phase error will be $\Delta\Omega T = 2\pi/5$ which is too much for acquisition. After a jump from $-1/20T$ to $1/20T$, we only have $\Delta\Omega T = \pi/5$ so the loop is capable of tracking the Döppler shift. The acquisition time for such a Döppler frequency jump is lower or equal to 10^4 for signal-to-noise ratios above 6 dB. That corresponds to 0.1 s if the data rate is 100 kbauds.

Classical Feedback Loop, Timing Jitter Influence

We will plot the root-mean square (rms) of the Döppler estimator, $\Delta\omega'$, as a function of the SNR expressed in dB.

$$\text{rms} = \sqrt{E[(\Delta\omega'T - E[\Delta\omega'T])^2]} \tag{3.35}$$

Figure 3.23 shows the rms as a function of the signal-to-noise ratio for different timing jitter values with a Döppler shift of $1/10T$. They show that the method is sensitive to timing jitter because of ISI due to Nyquist filters mismatch. The rms remains small and timing jitter does not cause any noticeable degradation on the bit error rate. The optimal timing instant should be searched as a function of the Döppler frequency shift.

QDPSK BER Performance

The system performance in terms of bit error rate will be particularly sensitive to the choice of filters in the receiver – both the BW filter and the LPF in the Döppler estimator branch.

The analytical expression of the average bit error probability for QDPSK with corrected Döppler is given in [15]. We do not repeat the expression here, only that the error probability expression includes the LPF gain factor χ, through the factor:

$$K_\chi = (1 + \chi)/(1 - \chi) \tag{3.36}$$

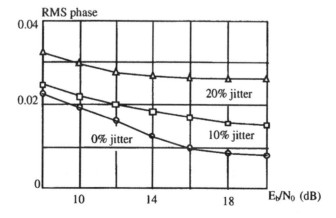

Figure 3.23 RMS phase vs. SNR classical feedback loop

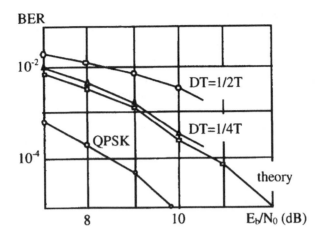

Figure 3.24   BER vs. E_b/N_0 for different Döppler shifts with QDPSK modulation.

This bit error probability is our reference and is named 'theory' in our figure.

In Figure 3.24, BER is plotted as a function of the SNR E_b/N_0 expressed in dB for different (constant) Döppler shifts (DS) when B (the BW filter bandwidth) stays matched with the transmit filter. The simulation and analytical results for DS = 0 were found to be in excellent agreement. It was observed that degradations due to Döppler shift are negligible (below 0.5 dB) up to a value of $1/4T$, but correspond to a 2 dB penalty for DS = $1/2T$.

We investigated the influence of widening B to $3/T$ for DS = $1/2T$ and DS = $3/4T$. We noticed that, for DS = $1/2T$, degradations caused by extra noise in the widened B were similar to those caused by the filter mismatch when the bandwidth was not widened.

In other words, if we do not accept any degradation above 0.5 dB (in comparison with 'theory'), a Döppler shift up to about $1/4T$ may be tolerated. If a 1 dB penalty is tolerated, we can consider a $3/8T$ Döppler shift with a brickwall bandwidth matched with the transmit filter. Furthermore, the results show that Döppler shifts up to half the data rate may be corrected at the expense of about 2 dB degradation. It is necessary to consider a multibeam division for Döppler shifts exceeding these orders of magnitude, or if the induced degradation is not accepted.

QDPSK Acquisition Range and Time
We have already seen that the acquisition range is limited by the BER degradation permitted. We will now discuss the influence of the LPF, which limits the acquisition range to some extent, and also the acquisition time due to the feedback filter design.

If the Döppler shift varies from $1/2T$ to $-1/2T$ on T_1 s (corresponding to 4×10^7 symbols in our numerical example), calculation leads to the following condition on the acquisition range:

$$B_1 > 2/(kT_1) \tag{3.37}$$

where B_1 is the half-bandwidth of the LPF and the DS varies from $1/kT$ to $-1/kT$. In this example, $2.5 \times 10^{-8} < 1 - \chi$ should be verified, which is the case with the chosen values of χ.

Simulation also indicates that this relatively slow Döppler variation does not cause any

BER degradation. Hence, the error rate depends only on the extreme values of DS, which means that if a 2 dB degradation is not acceptable, the beam must be divided.

When a beam is divided into two, the slow Döppler shift still has no influence on the performance, but the effect of the frequency jump when beams are changed must be studied. The LPF parameter χ plays an important role. The closer χ is to 1, the longer is the acquisition time. On the other hand, if we decrease χ in order to improve the acquisition range, we lose in BER performance because of increasing noise.

In order to evaluate the acquisition time analytically, the simplified situation is illustrated, where the evolution of the Döppler frequency shift as a constant in time just before and after the jump from $-\Delta\omega_1$ to $\Delta\omega_1$. A serial development of the output of the LPF as a function of the input gives us the following expression for the number of symbols, n, required for acquisition:

$$n = \log(0.293)/\log\chi \tag{3.38}$$

for DS $= 1/4T$.

Table 3.3 shows the acquisition time in number of symbols and that the analysis is confirmed by simulation. If $\chi = 0.99$ is chosen, about 120 symbols are lost after the handover. For a data rate of 100 kbauds, the acquisition time is about 1.2 ms. The noise in the feedback loop causes an additional 0.5 dB penalty on the BER in comparison with the theoretical QDPSK performance at a BER of 10^{-3}.

Table 3.3 Number of symbols needed for acquisition as a function of the LPF parameter χ

χ	BER penalty (dB)	Acq. time analysis	Acq. time simulation
0.95	1.25	25	24
0.99	0.5	123	122
0.995	0.45	250	244
0.999	0.4	1237	1227
0.9995	0.35	2460	2454

An identical analysis was performed for a jump from $-1/8T$ to $1/8T$, giving:

$$n = \log(0.5)/\log(\chi) \tag{3.39}$$

Note that the LPF design of the Döppler correction estimator has an important influence on the acquisition time.

QDPSK, Timing Jitter Influence

Due to the condition imposed on the timing instants $(\theta(k + 1/4)T) = \theta((k + 3/4)T))$ in order to obtain ISI-free symbols after the pseudo-matched filter, it is clear that the algorithm will be vulnerable to timing error in the receiver.

The rms of the Döppler estimation error was plotted, as in [15]. The analytical expression of the Döppler estimation error without timing jitter is also given in [15] (with $K\chi = (1 + \chi)/(1 - \chi)$ as a parameter) as:

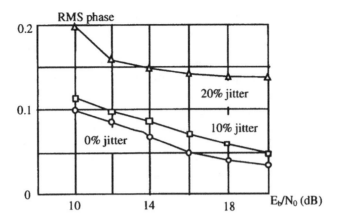

Figure 3.25 RMS phase vs. SNR for QDPSK with DS $= 1/10T$, B $= 2/T$ and $K\chi = 40$

$$\sigma^2_{\Delta\omega'T} = 4(1 + N_0 B/2P)/(K_\chi P/(N_0 B)) \tag{3.40}$$

where B the bandwidth of the BW filter, P the average signal power and N_0 the noise SPD. The root of the variance of the estimator $\sigma^2_{\Delta\omega T}$ is compared with the rms values deduced from simulation. The results of our simulations were in good agreement with theory, and showed that the rms of the estimator increases with the Döppler shift. Figure 3.25 shows that a timing jitter of about 10% can be tolerated without serious degradation of the estimator, but that when the timing jitter is about 20% the rms is multiplied by approximately two.

Simulations giving the BER for $\chi = 0.99$ and DS $= 1/4T$, were performed with different values of the timing jitter. They showed that a 10% timing jitter causes a degradation of about 0.5 dB on the BER and a 20% timing jitter provokes a BER floor at 10^{-2}.

QDDPSK BER Performance

An exact analytical expression of the average bit error probability in the absence of frequency error is not available for a QDDPSK, but an upper bound is given in [17] both for the (T,T) and the $(T,2T)$ implementation. The analytical expression is valid for $\alpha = 0$ and when the bandwidth of the brick-wall filter is $B = 1/T$. This upper bound will be the reference (in terms of dB lost) for our simulations.

First we consider a transmission without Döppler shift, without the BW filter and with a Nyquist filter evenly distributed between the transmitter and the receiver. The roll-off factor is $\alpha = 0$. Plots of the BER as a function of the SNR E_b/N_0 expressed in dB for the (T,T) and the $(T,2T)$ algorithms showed good agreement between simulation and the analytical results. They also showed that more than a 1.5 dB gain is achieved with the $(T,2T)$ over the (T,T) implementation at a BER of 10^{-4}.

We then considered an implementation with a Nyquist transmit filter and a BW filter at the receiver. The BER as a function of the SNR was investigated for different values of the Nyquist roll-off factor α and without Döppler shift. The BW filter bandwidth was matched with the transmit filter, i.e. $B = (1 + \alpha)/2T$. The simulations indicated that we lose 0.5 dB for $\alpha = 0.1$ in comparison with the distributed Nyquist filter with $\alpha = 0$. Due to increasing noise in the double differential detector, degradations caused by increasing α are important; we lose about 3 dB if α varies from 0 to 1, for both implementations.

Further simulations on the $(T, 2T)$ implementation showed that the adapted BW-filter in combination with small values of α gave poor results – not even a Döppler shift of $1/8T$ was corrected for $\alpha = 0.1$. For $\alpha = 0.3$, $1/8T$ was corrected, but not $1/4T$. By widening the BW filter by an amount corresponding to the Döppler shift, signal acquisition was possible for $\alpha = 0.1$ at DS $= 1/8T$. However, this caused some degradation (about 1.5 dB additionally) due to the increase of the noise. For $\alpha = 0.3$, the degradation caused by widening B is similar to the effect of filter mismatch. For DS $= 1/4T$ an additional 3 dB loss must be expected either from mismatch or noise. All degradations referred to correspond to a BER $= 10^{-3}$.

QDDPSK Acquisition Range and Time
With relatively slow Döppler shift variation considered in this paper, the BER depends on the extreme values of DS. As indicated previously, a Döppler shift of $1/2T$ is too much for this algorithm; a division into spotbeams is necessary. The acquisition range seems to be limited to about $1/4T$ for the chosen filter parameters. On the other hand, simulations indicated a very quick re-acquisition after frequency jump; less than 10 symbols are lost. This is due to the fact that no feedback loop is implemented in the receiver and that the estimation of the Döppler shift is completed after only three symbol intervals (with the $T, 2T$ implementation). Thus, the only criterion that has to be taken in account in the choice of the number of beams, is the extreme value of the Döppler shift for the BER performance.

QDDPSK, Timing Jitter Influence
Since $n(t)$ and $n(t - T)$ are uncorrelated, the analytical expression of the rms of the Döppler estimator is simply:

$$\text{rms} = \sqrt{N_0 B / 2} \qquad (3.41)$$

where N_0 is the noise SPD and B the BW filter bandwidth. The correctness of the simulated rms values was checked by comparison with the analytical expression for 0% timing jitter. Figure 3.26 shows the rms Döppler estimation error when DS $= 1/16T$ for $\alpha = 0.3$ and B not

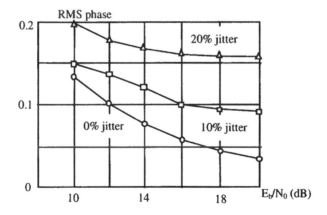

Figure 3.26   RMS vs. SNR for QDDPSK with $B = (1 + \alpha)/T$, $\alpha = 0.3$ and DS $= 1/16T$

widened for 0%, 10% and 20% timing jitter. Simulations show that this algorithm is quite sensitive to timing jitter.

3.4 Catching Co-Channel Interference

This section starts with a description of the model used in this study, which as previously mentioned, was first introduced by Jamalipour et al. [18]. A brief introduction to the IPhP3 is given follows, comparing some of its relevant attractive features with the commonly used Poisson process on the plane. In the next section, the first two moments of the cumulated interference power are discussed. Our aim in this section is to highlight the powerful technique associated to the IPhP3. Readers interested in deep mathematical details are recommended to read Remiche [20].

3.4.1 Satellite System Model

In this section, we are interested in modelling a LEO satellite communication system based on a CDMA scheme. According to the traffic distribution over a geographical area, the total interference power for the uplink can be derived and its two first moments analysed.

LEO satellites are non-geostationary satellites. The area which they cover on Earth, while being of fixed shape, is moving over time. LEO satellites are located at a fixed radius h from the Earth surface. The number N_S of satellites present on a same orbital is also fixed. The following basic features of a LEO satellite system are defined. The *coverage area* of a LEO satellite system is specified as the set of points for which the elevation of the satellite over the horizon is big enough; the acceptable minimum elevation angle θ is designed by the system. In contrast, the *interference area* is determined by the final line of sight of the satellite. Any Earth station located in the interference area, but out of the coverage area of some satellite is not allowed to connect to this satellite even if its signal reaches the satellite. The area covered by two satellites at the same time is called the *double coverage area*. In this work, we assume it does not exist and consider only two satellites. In our model, the coverage and interference area of Satellite 1 and the coverage area of Satellite 2 are approximated via circles. The distance d between the centres of coverage areas is equal to:

$$d = R\sin(2\pi/N_S) \tag{3.42}$$

where R is the Earth radius and the satellites are assumed to be evenly spaced in the orbits. Thus the radius r of the coverage area circle limit is equal to:

$$r = d/2 \tag{3.43}$$

The radius R_I of the interference area is as in Jamalipour et al. [18], specified via the interference limit angle β_I, where

$$\beta_I = \arccos R/(R+h) \tag{3.44}$$

so that we approximate R_I as:

$$R_I = R\sin\beta_I \tag{3.45}$$

In a CDMA system, all the users are sharing a common bandwidth. Transmission is made possible through signal coding and power control mechanisms. If only two satellites are considered, two kinds of coded messages received at Satellite 1 exist. First, there are the coded messages coming from its covered area. Those are received at the required power S_1. Secondly, for Earth stations which are located in the coverage area of Satellite 2, the messages are received at Satellite 1 with the power $S(\rho,\theta,S_2)$, depending on their location (ρ,θ) and the power S_2 required by their own communicating satellite (Satellite 2). We assume here that the received power $S(\rho,\theta,S_2)$ is equal to:

$$S(\rho,\theta,S_2) = S_2 \tag{3.46}$$

where $d^2(\text{Satellite } i, \rho, \theta)$ is the square of the distance between Satellite i, $i \in \{1,2\}$ and Earth station located at (ρ, θ). Indeed in [18], the authors assume that shadowing and Raleigh fading have relatively small effect, so that the free propagation model can be used. If the required received power at Satellite 2 is S_2, then an Earth station located at (ρ, θ) must transmit its signal with the power $S_2 d^2(\text{Satellite } 2,\rho,\theta)$. Only Earth stations located in both the interference area of Satellite 1 and the coverage area of Satellite 2 can be received by both satellites.

We are interested in the total interference power received at Satellite 1. This is a key parameter in describing the quality of signals received at the satellite from Earth stations. Indeed, this quality is often measured in terms of the signal-to-interference (SIR) ratio, i.e. the ratio $\text{SIR} = S_1/I$, where S_1 is the received signal power of the Earth station at Satellite 1 and I is the total interference power received at the satellite.

The location of the Earth stations in the geographical area of interest is directed by an IPhP3, which is in the follwong section.

3.4.2 IPhP3 with Deterministic Marks

Some of the main characteristics of IPhP3 are introduced and explained, followed by the marked process related to the model presented in the previous section.

Let N be an IPhP3, that is a set of points randomly located on the plane via their polar coordinates, i.e. $\{(\rho_n, \theta_n); n \in IN_0\}$, with $\rho_n \in IR^+$ and $\theta_n \in [0, 2\pi)$. One defines, for all $n \in IN_0$:

$$\rho_n = \sqrt{T_n/\pi} \tag{3.47}$$

where $\{T_n; n \in IN_0\}$ are the random points generated by a stationary MAP on the positive real axis. A MAP $\{(L(t), Y(t)); t \in IR^+\}$ is a Markov process defined on the bi-dimensional state-space $INx\{1,...,m\}$ with generator:

$$Q = \begin{pmatrix} D_0 & D_1 & 0 & \Lambda \\ 0 & D_0 & D_1 & \Lambda \\ 0 & 0 & D_0 & \Lambda \\ M & M & M & O \end{pmatrix} \tag{3.48}$$

where the matrices D_0 and D_1 are two square matrices of size m (with m finite) where negative elements are exclusively present on the diagonal of D_0 and

$$(D_0 + D_1)\underline{1} = \underline{0} \tag{3.49}$$

The vector $\underline{\delta}$ is the stationary vector of $D_0 + D_1$:

$$\underline{\delta}(D_0 + D_1) = 0 \tag{3.50}$$

$$\underline{\delta}\underline{1} = 1 \tag{3.51}$$

The first dimension $L(t)$ of the process is called the *level* of the process, while $Y(t)$ is called the phase. The matrices D_0 and D_1, respectively, represent the transition rate among phases when the process stays in the same level or goes up one level, i.e. $(D_0)_{ij}$ is the instantaneous rate of change from (n,i) to (n,j) and $(D_1)_{ij}$ is the instantaneous transition rate from (n,i) to $(n + 1,j)$, for $n \geq 0$, $1 \leq i \neq j \leq m$. No transition may occur in one step from (n,i) to $(n + k,j)$ with $k \geq 2$ or from (n,i) to $(n - k,j)$, for $k \geq 1$. The epochs $\{T_n; n \in IN_0\}$ where a change of level occurs constitute a set of random locations of points on the positive real axis.

Secondly, the set of angular coordinates $\{\theta_n; n \in IN_0\}$ is composed of independent and uniformly distributed random variables over $[0, 2\pi)$. Those variables are also independent of the MAP process directing the values of the radial coordinates.

The IPhP3 family includes the Poisson process as a particular case. IPhP3's are similar to the Poisson process in that their intensity measure is a constant multiplied by the area of the set of interest:

$$E[N(B)] = \underline{\delta}D_1\underline{1}|B| \tag{3.52}$$

Where B is a set of the plane and $N(B) = (N \cap B)$ and $|B|$ is the area of the set B. Contrary to the Poisson process, IPhP3's allows patterns of points where first higher moments of counts are not direct functions of the area. Moreover, dependence between numbers of points in disjoint sets is possible to incorporate into models.

In this context, each point (ρ, θ) belonging to N is assigned a mark: $m(\rho, \theta)$. This mark is the value of a deterministic real-valued function $m(.)$ defined on the plane. In this work, a point on the plane represents the random location of an Earth station, while its mark represents the power at which the Earth station is received at Satellite 1. Since on average, the number of Earth stations in any of the two coverage area will be the same, it is assumed that the required power at any two satellites is fixed and equal to S (see Ref. [22]). In this model, it suggests that Earth stations located in the coverage area of Satellite 1 are marked by the value S and those located in both interference area of Satellite 1 and coverage area of Satellite 2 are marked with the value $S(\rho, \theta, S)$ where the function $S(.)$ is defined as in Eq. (3.1).

3.4.3 Two First Moments of Cumulated Interference Power

As previously explained, the total interference power P is composed of two parts:

$$P = P_I + P_E \tag{3.53}$$

where P_I is the total intra-cell interference power due to Earth stations situated in the coverage area of Satellite 1, and P_E is the total extra-cell interference power generated by Earth stations located in both the coverage area of Satellite 2 and the interference area of Satellite 1.

We begin with the computation of the mean total interference power received at Satellite 1:

$$E[P] = E[P_I] + E[P_E] \tag{3.54}$$

Since we expect $\underline{\delta}D_1\underline{1}\ \pi r^2$ points to be located in the coverage area of Satellite 1 (see Eq.

(3.2)) and that each of those Earth stations is received with a power S, it can be concluded that:

$$E[P_1] = S\underline{\delta}D_1\underline{1}\pi r^2 \tag{3.55}$$

The mean of the extra-cell interference power is rather complicated. We explain the main idea of its derivation without going too far into the mathematical details.

Let us first define $P_E(x)$, for $x \in [\pi r^2, \pi r_I^2)$: the total interference power from Earth stations located in the intersection of the coverage area of Satellite 2 and a subset of the interference area of Satellite 1, precisely the set:

$$C(x) = C_x(0,0) \cap C_r(2r,0) \tag{3.56}$$

where $C_x(0,0)$ is a circle of radius x and whose centre is located at $(0,0)$. Clearly, we have $P_E = P_E(\pi r_I^2)$. We also define the matrix $\mu^{E*}(x)$, with $x \in [\pi r^2, \pi r_I^2)$ as being a matrix where its ijth element is:

$$\mu_{ij}^{E*}(x) = E[P_E(x)1[Y(\pi r^2) = j]|Y(x) = i] \tag{3.57}$$

with i, j are in $\{1,...,m\}$. It is obvious that:

$$\underline{\delta}\mu^{E*}(\pi r_I^2)\underline{1} = E[P_E] \tag{3.58}$$

In order to determine $E[P_E]$, we focus on the first radial location of Earth station observed in the interval $[\pi r^2, x]$, starting form the right. The radial location of the first Earth station on the left of x is y with probability $\exp\{D_0^*(x - y)\} D_1^* dx$ (see Ref. [21]). The matrices D_0^* and D_1^* are defined as:

$$D_0^* = \operatorname{diag}(\underline{\delta})^{-1}D_0^T\operatorname{diag}(\underline{\delta}) \tag{3.59}$$

$$D_1^* = \operatorname{diag}(\underline{\delta})^{-1}D_1^T\operatorname{diag}(\underline{\delta}) \tag{3.60}$$

where $\operatorname{diag}(\underline{\delta})$ is a diagonal matrix with $(\operatorname{diag}(\underline{\delta}))_{ii} = \delta_i$. They represent the reversed-direction MAP process, intuitively the MAP process determined on the MAP in reversed sense. Using this approach, we get:

$$\mu^{E*}(x) = \int_{\pi r^2}^{x} \exp D_0^*(x - y))D_1^* dx\{\mu^{E*}(y) + \exp\{(D_0^* + D_1^*)y\} \int_{-\theta(y)/2}^{\theta(y)/2} \frac{d\theta}{2\pi} S(\sqrt{y/\pi}, \theta, S)\} \tag{3.61}$$

where the function $\theta(y)$ gives the amplitude of the intercepted angle at location y of the circle $C_r(2r,0)$, that is:

$$\theta(x) = 2\arccos\frac{x - 3\pi r^2}{4r\sqrt{\pi x}} \tag{6.62}$$

By simple differentiation of this last expression with respect to x, we get:

$$\partial\mu^{E*}(x) = (D_0^* + D_1^*)\mu^{E*}(x) + D_1^*\exp\{(D_0^* + D_1^*)x\} \int_{-\theta(x)/2}^{\theta(x)/2} \frac{d\theta}{2\pi} S(\sqrt{x/\pi}, \theta, S) \tag{3.63}$$

By pre-and post-multiplying this expression by $\underline{\delta}$ and $\underline{1}$, and solving this differential equation, we obtain:

$$E[P_E] = S\underline{\delta}D_1\underline{1} \int_{\pi r^2}^{\pi r_i^2} dx \left(\theta(x) + \frac{\theta(x) - 4\sqrt{x/\pi}\sin\theta(x)/2}{x/\pi + h^2} \right) \tag{3.64}$$

So that:

$$E[P] = S\underline{\delta}D_1\underline{1}\pi r^2 + S\underline{\delta}D_1\underline{1} \int_{\pi r^2}^{\pi r_i^2} dx \left(\theta(x) + \frac{\theta(x) - 4\sqrt{x/\pi}\sin\theta(x)/2}{x/\pi + h^2} \right) \tag{3.65}$$

The next step is the second moment computation. One has that:

$$E[P^2] = E[P_I^2] + E[P_E^2] + 2E[P_I P_E] \tag{3.66}$$

Let us first define the function $P_I(x)$, for $x \in [0, \pi r^2]$, as the cumulated interference power coming from area $C_x(0,0)$. Next we define $\mu^{I*}(x)$, with $x \in [0, \pi r^2]$, as a matrix composed of the functions

$$\mu_{ij}^{I*}(x)(i,j \in \{1,...m\} \tag{3.67}$$

where

$$\mu_{ij}^{I*}(x) = E[P_I(x) \; 1[Y(\pi r^2) = j]|Y(x) = i]. \tag{3.68}$$

The random variables P_I and P_E are dependent through the phase of MAP process at location πr^2 only, since it affects the radial location of the points both in $C_r(0,0)$ and $C_r(2r,0)$. Conditioning on the phase, we get:

$$E[P_I P_E] = \sum_{1 \le i \le m} E[P_I 1[Y(\pi r^2) = i]]E[P_E|Y(\pi r^2) = i] = \underline{\delta}\mu^{I*}(\pi r^2)\mu^{E*}(\pi r_i^2)\underline{1} \tag{3.69}$$

The function $\mu^{I*}(x)$ can be determined using the same approach as the one used to obtain the function $\mu^{E*}(x)$. In order to compute the two other terms in Eq. (3), we define the matrices $\mu^{I2*}(x)$ and $\mu^{E2*}(x)$ in the same way as for $\mu^{I*}(x)$ and $\mu^{I*}(x)$, respectively, where this time the functions P_I and P_E are elevated to the square power. We define the system that has to be numerically solved in order to get the two first moments of the cumulated interference power observed at Satellite 1. Let $\mu^{X*}(x)$, with $X = I$ or E be a column-vector, defined as:

$$\bar{\mu}^{X*}(x) = \begin{pmatrix} \mu^{X*}(x)\underline{1} \\ \mu^{X2*}(x)\underline{1} \end{pmatrix} \tag{3.70}$$

and let $A(x)$ be a matrix composed of:

$$A(x) = \begin{pmatrix} D_0^* + D_1^* & 0 \\ 2D_1^* Sf_X(x) & D_0^* + D_1^* \end{pmatrix} \tag{3.71}$$

and let $b(x)$ be a column-vector defined as:

$$b(x) = \begin{pmatrix} D_1^*\underline{1}Sf_X(x) \\ D_1\underline{1}S^2 f_X^2(x) \end{pmatrix} \tag{3.72}$$

where $f_X(x)$ is a real-valued function with:

$$f_1(x) = 1, \text{ for } x \in [0, \pi r^2)$$ (3.73)

$$f_E(x) = (\theta(x) + 2(\theta(x) - 4\sqrt{x/\pi} \sin\theta(x)/2)/(x/\pi + h^2)), \text{ for } x \in [\pi r^2, \pi r_1^2)$$ (3.74)

The first two moments of the cumulated interference power received at Satellite 1 and coming either exclusively from the coverage area of Satellite 1 or from its interference area is obtained by solving the following system:

$$\partial \bar{\mu}^{X^*}(x) = A(x)\bar{\mu}^{X^*}(x) + b(x)$$ (3.75)

with initial condition $\underline{\mu}^{X^*}(0) = 0$ and integrated on the interval $[0, \pi r^2]$ if $X = I$, or $[\pi r^2, \pi r_1^2]$ otherwise. This system can be obtained using the technique previously highlighted: first conditioning on the first right radial coordinate and then deriving the expression according to x.

The interest in such a mark point process is that it can be used in any model where the exact location of users or Earth stations must be known, and any parameters which are related to these locations.

3.5 Chapter Summary and Perspectives

We have given an overview of the most common candidates for modulation and channel coding for satellite communication and have discussed each of the schemes according to the requirements of a packet switched system with multimedia applications.

We concentrated on a modulation scheme called variable rate N-MSK. The variable rate N-MSK modulation technique was described and analysed in non-terrestrial broadband communication systems, where the power consumption is limited. The chosen modulation schemes have good performance if the non-linear amplifier as high power amplifier is used. The eye closure switching criteria was chosen to switch between 1-MSK and 2-MSK modulation schemes. A slow time varying radio channel was used to simulate the effects of time varying channel attenuation. The channel attenuation obeys log-normal distribution. If the switching level is chosen in coherence with the channel characteristics, the average transmitted bits per symbol is increased and BER stays within the required limits. The proposed variable modulation scheme is especially suitable for communication systems, which support different services for different users. If the higher data rate is required from the user and required BER and delay can be achieved, the system switches to a more bandwidth efficient modulation scheme. If more than two modulation schemes are used in communication system, an improvement in average number of transmitted bits per symbol is expected.

As the need for higher bandwidth efficiency in satellite systems increases, the use of higher modulation schemes may be required. In that case, efficient techniques for predistortion will be implemented. Many satellite operators are currently investigating and even operating satellite links based on 16 QAM.

We suggested one particular coding scheme that complies with requirements of efficiency and packet transmission suitability – the TPC. This code is adapted to packet transmission since each packet may be considered a codeword and treated separately for decoding. Due to the Turbo code principle with soft input/soft output and iterative decoding, the performance achieved with these codes is about 3 dB from the Shannon limit at 10^{-5} after only 4 matrix decoder iterations. The delay involved in decoding may be reduced to only one packet delay if the bit rate is low enough to permit the use of a system clock performing multiple iterations. If the number of decoder iterations necessary is four, one would need a decoder working eight times as fast as the bit rate in order to finish decoding within one packet duration.

The channel coding performance obtained with Turbo codes is continuously enhanced by the introduction of new Turbo codes. Recently, a new coding scheme proposed by Berrou [11] promises better performance results than the TPC, with comparable decoder complexity. These new codes are also well adapted to the type of packet transmission we are considering in this paper.

As a summary, we have presented some of the most common coding and modulation schemes in Figure 3.27. The representation of number of bits transmitted per symbol (close to the spectral efficiency, a correction factor due to the Nyquist roll-off is necessary) as a function of the SNR E_b/N_0, permits a comparison between schemes and an indication on the closeness to the Shannon limit. CPM are not presented in the figure; the infinite bandwidth character of the CPM, makes it difficult to compare them with finite bandwidth modulation schemes like PSK and QAM.

As a conclusion to the synchronisation study, we will compare the three methods with the following choice of parameters: for the differential detection, QDPSK, we choose $\chi = 0.99$ and a BW filter not widened with a Döppler shift of $1/4T$. The $(T,2T)$ implementation is chosen for the double differential detection, QDDPSK, with a roll-off factor of the Nyquist filter $\alpha = 0.3$. The same value of α is supposed for the classical feedback loop acquisition method.

The feedback loop technique has a much better performance in terms of BER than the two forward loop methods without Döppler shift. We lose at least 2.5 dB with the QDPSK technique and at least 5 dB with the QDDPSK for a BER $= 10^{-4}$ without Döppler shift.

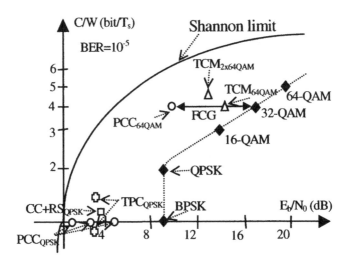

Figure 3.27 Spectral efficiency as a function of the SNR for a BER $= 10^{-5}$. The Shannon limit and some unmodulated linear PSK and QAM schemes are shown, as well as examples of coded schemes. The concept of Fundamental Coding Gain (FCG) is also shown

The bit error rate depends to a large extent on the Döppler shift value for the classical feedback loop technique; whereas Döppler shift of $1/4T$ for the QDPSK and $1/8T$ for the QDDPSK can be considered without serious BER degradations (less than 0.5 dB) when compared to the performance without Döppler shift. For all three methods, the filter mismatch due to the Döppler shift will degrade the performance results. It is possible to compensate for this mismatch in several ways, but all solutions lead to other problems. The best method is to find a compromise in the system characteristics and required performance goals.

When a Döppler shift varies from $1/2T$ to $-1/2T$ in 4×10^7 symbol periods, only the QDPSK algorithm may be employed without beam division. The classical feedback loop can only correct Döppler shifts up to about $1/10T$ and the QDDPSK up to about $1/8T$. The BER performance is not degraded with varying Döppler shift, but the Döppler frequency jump after handover influences the performance. Only frequency jumps from $-1/20T$ to $1/20T$ can be tolerated by the feedback loop technique with a re-acquisition time about 10,000 symbols. The re-acquisition time after handover is almost immediate for the QDDPSK and of about 100 symbols for the QDPSK with the LPF parameter $\chi = 0.99$.

The rms of the different estimations confirms that, once locked and without timing jitter, the feedback technique performs better than the two other methods. In terms of BER degradation, a timing jitter of 20% always causes important additional degradation (more than 4 dB for the feedback loop and the QDPSK and more than 6 dB for the QDDPSK). The QDPSK is quite resistant to a timing jitter of 10% with a loss of less than 0.5 dB; whereas the same timing jitter causes a loss of more than 1 dB for the feedback loop and more than 2.5 dB for the DDPSK. The very simple design of the QDDPSK algorithm must be noted as favourable for implementation.

Finally, one section was dedicated to the analysis of the cumulated interference power at a satellite in a LEO satellites telecommunication system. The model was restricted to the study of two satellite footprints, and the assumed access scheme was CDMA. The method presented was the IPHP3 process, which is a set of point processes in the plane. In this context, the first two moments of the cumulated interference power at one satellite were derived. Further work of interest would extend the model to more than two satellites. Such an extension would prove difficult for the calculation of the second moment, whereas the calculation of the first moment is expected to be quite straightforward. Another area of interest would be to study the effect of a double coverage. Again, the complexity of the calculation could prove to be prohibitive in this exercise.

References

[1] J.G. Proakis, *Digital Communications*, 2nd Edition, New York, McGraw-Hill, 1989.

[2] G. Karam and H. Sari, 'Generalized data predistortion using intersymbol interpolation', *Philips Journal of Research*, Vol. 46, No. 1, 1991.

[3] J.B. Anderson, T. Aulin and C.-E. Sundberg, *Digital Phase Modulation*, New York, Plenum Press, 1986.

[4] H. Sari, G. Karam and V. Paxal, 'Trellis-coded constant-envelope modulations with linear receivers', *IEEE Transactions on Commununications*, Vol. 44, No. 10, 1996.

[5] G.D. Forney, 'Coset codes – Part 1: introduction and geometrical classification', *IEEE Transactions Information Theory*, Vol. 34, No. 5, 1988.

[6] A.M. Michelsen and A.H. Levesque, *Error-control Techniques for Digital Communication*, New York, Wiley, 1985.

[7] ETS 300 421, 'Digital broadcasting systems for television, sound and data services – framing structure, channel coding and modulation for 11/12 GHz satellite services', *ETSI draft*, August 1994.

[8] G. Ungerboeck, 'Channel coding with multi-level/phase signals', *IEEE Transactions Information Theory*, Vol. 28, No. 1, 1982.

[9] C. Berrou, A. Glavieux and P. Thitimajshima, 'Near Shannon limit error-correcting coding and decoding: turbo codes (1)', *IEEE International Conference on Communication ICC '93*. Vol. 2/3, 1993.

[10] R.M. Pyndiah, 'Near-optimum decoding of product codes: block turbo codes', *IEEE Transactions on Communication*, Vol. 46, No. 8, 1998.

[11] 'Chip description of Turbo Product Codes, version 3.1', June 1998, Efficient Channel Coding, Inc., www.eccincorp.com.

[12] S. Benedeto, E. Biglieri and V. Castellani, *Digital Transmission Theory*, London, Prentice Hall, 1987.

[13] R. Steele, *Mobile Radio Communications*, London, Pentech Press, 1992.

[14] G.M. Djukic and J. Freidenfelds, 'Establishing wireless communications services via high-altitude aeronautical platforms: a concept whose time has come?' *IEEE Communication Magazine*, Vol. 35 No. 9, pp. 128–135, 1997.

[15] T. Javornik and G. Kandus, 'Variable rate CPFSK modulation technique', V: Ruggieri, Marina (ur.), Mobile and Personal Satellite Communications 3. *Proceedings of the Third European Workshop on Mobile/Personal Satcoms (EMPS '98)*, Venice, November, 1998, London, Springer-Verlag, 1998, pp. 376–388.

[16] M.K. Simon and D. Divsalar, 'Döppler corrected differential detection of MPSK', *IEEE Tansactions on Communications*, Vol. 37, No. 2, pp. 99–109, 1989.

[17] T.C. Jeffrey, E.H. Satorius and M.J. Agan, 'A frequency offset estimation and compensation algorithm for K/Ka-band communications', *International Journal of Satellite Communications*, Vol. 14, pp. 191–200, 1996.

[18] M.J. Miller, B. Vucetic and L. Berry, *Satellite Communications, Mobile and Fixed Services*, Kluwer Academic, 1993.

[19] A. Jamalipour, M. Katayama, T. Yamazato and A. Ogawa, 'Signal-to-interference ratio of CDMA in low Earth-orbital satellite communication systems with non-uniform traffic distribution', *Proceedings of the IEEE Globecom Conference*, San Francisco, CA, pp. 1748–1752, 1994.

[20] G. Latouche and V. Ramaswami, 'Spatial point processes of phase-type'. in V. Ramaswami and P. Wirth, (Editors), *Teletraffic Contribution for the Information Age*, pp. 381–390, Amsterdam, Elsevier, 1997.

[21] M.-A. Remiche, Phase-type planar point processes, *Analysis and Application to Mobile Telecommunications Performance Studies*, PhD thesis, Université Libre de Bruxelles.

4

Networking

Over the past 30 years, geostationary satellites have been used almost exclusively to provide commercial satellite communications. They will continue to play an important role particularly for broadcast applications. However, the requirements for high-power terminals and the long propagation delay caused by the high altitude of the satellites implies that the quality of service for many internet applications provided via geostationary satellites will be degraded significantly. The recent development of non-GEO satellites to counteract the deficiency of geostationary satellites over the last decade has enabled the launch of satellite personal communication services.

Non-GEO satellite systems can help to meet the demand for information by providing global access to the telecommunications infrastructure, currently available only in advanced urban areas of the developed countries. Like GEO satellites, non-GEO satellites also provide a means of interconnecting LAN on a global basis. The main advantage of the deployment of non-GEO satellites is their relatively low altitudes in comparison with GEO satellites. This enables non-GEO satellites, especially LEO satellites, to provide services with delay which is compatible, or nearly compatible, with that introduced by the terrestrial networks. Because of their lower altitudes, lower power terminals are possible. This in turn has a positive effect in the size and cost of the terminals. The evolution from GEO to LEO satellites has resulted in a number of global satellite systems such as the Iridium, which is now in the sky, the Globalstar and ICO.

However, non-GEO satellite systems impose different challenges in network design. Firstly, the topology of non-GEO satellite-based networks is dynamic due to the satellite motion. The satellite position and propagation delay relative to satellites in the other planes and to the terminals changes continuously but in a predictive manner. Handover between satellites may be required in order to ensure service continuity. Furthermore, traffic queues which accumulate in the satellites may also change the waiting time before transmission to the next satellite. All these factors affect the resource control mechanisms adopted in the network, such as the routing strategy and congestion control techniques. Furthermore, the combination of low altitude and high elevation angles results in a small footprint for each satellite and hence large numbers of satellites are required for global coverage. To support the network reliability requirements, a high degree of coverage redundancy and the use of on-orbit spares will have to be supported.

In this chapter, several network control issues are considered. Section 4.2 investigates issues in the use of non-geostationary to provide LAN interconnection. This is followed by detailed description on network resource control in Section 4.3. Under Section 4.3, aspects related to the routing strategy, call control functions and multicasting techniques are studied. Results for different studies are presented. Section 4.4 provides a detailed description on the reliability requirement and techniques for satellite systems. Finally in Section 4.5, the security requirements and their implementation issues are discussed.

4.1 LAN Interconnection

4.1.1 Introduction

COST226 (Integrated Space/Terrestrial Networks) has demonstrated the feasibility of LAN interconnection by GEO systems using transparent satellites and fixed low-cost VSAT Earth stations [1], COST253 continues the work on LAN interconnections, with non-GEO satellites. GEO systems work well for fixed, non-interactive services like direct television and broadcasting, whereas non-GEO solutions have received a great deal of attention for real-time and personal communication. The non-GEO satellite systems within the context of COST253 will compensate for the lack of sufficient terrestrial high-bit-rate links mainly by interconnecting regional or national distributed LANs. The advantage of using a non-GEO satellite constellation is that real-time services can be supported and that global coverage can be achieved with a higher availability. Recent proposals for a non-GEO broadband satellite constellation supporting portable/fixed terminals demonstrates the trend towards lower orbits than GEO, which has been used by all commercial communications satellites over the last 30 years.

In this section, the network architecture is described providing some detail about each network component. The general network architecture for a meshed-satellite network using On-Board Switching (OBS) satellites with Inter-Satellite Links (ISLs) is also described. The protocol stack for a scenario using Internet Protocol (IP) or ATM is illustrated, followed by a description on the Terrestrial/Satellite Network Termination Modules (TNTM/SNTM), which carry out functions such as protocol mapping, modulation/demodulation and channel coding/decoding. The modems and codecs will also be part of the satellite payload. However, modulation/demodulation and channel coding/decoding functions are not covered in this chapter.

Important design criteria such as coverage, availability, path delay and path loss are discussed. Factors contributing to the satellite availability are described in detail. The satellite payload architecture and the advantages/disadvantages of OBS and ISLs and a comparison of four different on-board switch architectures. Finally the conclusions are provided.

4.1.2 Satellite Network Architecture

In [3], recent non-GEO broadband proposals were compared in order to select the optimum satellite constellation and satellite architecture. Within COST253, OBS satellites which employ ISLs were selected to make routing in the sky possible without being dependent on the terrestrial infrastructure. Apart from LEO constellations like Celestri and Teledesic, hybrid constellations such as WEST and Spaceway were also of interest. Figure 4.1 shows the network architecture for the interconnection of LANs using a non-GEO satellite constellation. This architecture corresponds to reference architecture 2.3 of the Telecommunications Industries of America (TIA) Communications and Interoperability group [4].

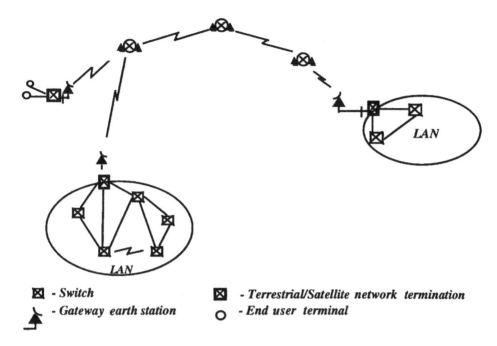

- Switch

- Gateway earth station

- Terrestrial/Satellite network termination

- End user terminal

Figure 4.1 Satellite network architecture

The ATM protocol layer stack for OBS satellites is shown in Figure 4.2. The application can either be transported directly over ATM or via TCP/IP over ATM. The use of standard ATM protocols to support seamless wired and wireless networking is possible by incorporating a new radio specific protocol sublayer into the ATM protocol model [5]. Considering that satellite communications use multiple access on a shared medium, a Medium Access Control (MAC) layer which is not present in traditional ATM networks, is needed. The MAC protocol plays a central role as means of accessing the Radio Physical Layer (RPL) from the ATM layer. The access scheme refers to the physical-layer multiplexing technique to share a common channel among multiple users of possibly multi-services.

The problem of statistical multiplexing at the satellite–air interface is slightly different to that in the fixed network. In the fixed network, the problem is associated with the control of bandwidth on an outgoing link from some multiplexing point after buffering has been performed. It can be implied that the access links from the source are dimensioned in such a way that they do not impose any constraints on the traffic (e.g. sources can transmit at their peak bit rate). In the air interface, the constraint is on the bandwidth available in total to all sources before the buffering/multiplexing point.

The satellite-user interface has to contain support for the mapping of user terminal connections to the shared satellite access link. A key issue is the mapping of service classes to the satellite channel, so as to maintain the required QoS for each virtual connection. A Logical Link Control (LLC) header to facilitate error recovery mechanisms is optional. A MAC protocol, designed to allow statistical multiplexing of ATM traffic over the satellite–air interface, especially in the uplink for the independent and spatially distributed terminals, has been investigated in [6,7].

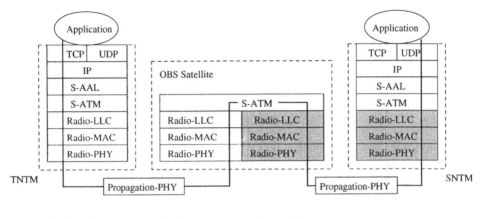

S-AAL:Satellite ATM Adaptation Layer
S-ATM:Satellite ATM Layer
   LLC:Logical Link Control Layer
   MAC:Medium Access Control

PHY:Physical Layer
OBS:On-Board-Switching
TNTM:Terrestrial Network Termination Module
SNTM:Satellite Network Termination Module

Figure 4.2   ATM protocol layer stack for OBS satellite

4.1.3 Terrestrial/Satellite Network Termination Module Characteristics

The functions of the SNTM are similar to the TNTM, but in the reverse order. In this section the functions of the TNTM, also applicable to the SNTM, are described. Figure 4.3 shows some of the main transmit functions in an Earth station. At the receiver, the equivalent reciprocal functions have to be performed.

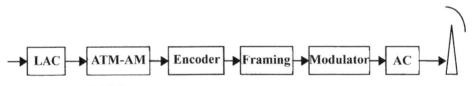

LAC: LAN ATM Converter
ATM-AM: ATM Adapter Module
AC: Access Control

Figure 4.3   Block diagram of ground station transmit functions

4.1.3.1 Protocol Mapping/Tunnelling

This module carries out mapping/tunnelling of the terrestrial network protocol to the satellite protocol. Two protocols have been selected to be used for transmission over satellite, namely IP and ATM. First the modules required for ATM transmission are explained and this is followed by the modules required for IP transmission over the satellite.

The LAN ATM Converter Module
The LAN ATM Converter (LAC) module converts the FDDI and Ethernet packets into ATM

cells, then passes the cells to the ATM Adaptation Module (ATM-AM). A large buffer may be required.

The ATM Adapter Module (ATM-AM)

The ATM-AM is an ATM Adapter. It multiplexes the ATM cell streams from the input ports into one ATM cell stream. For example in the RACEII CATALYST project [2] this module had buffers of 154 cells for port 0 and 77 cells for port 1. The output rate of the ATM-AM module is dependent on the maximum possible uplink rate of the satellite. It is anticipated that the satellite uplink bandwidth will not be sufficient to support the terrestrial ATM rates hence some buffering on the ground terminal is required. This module provides an interface between the terrestrial network and the satellite ground-station.

4.1.3.2 The IP Adapter Module (IP-AM)

The IP adapter module multiplexes the IP packet streams from the input ports into one IP packet stream. Similar to the ATM-AM the output rate of the IP-AM module is dependent on the maximum possible uplink rate of the satellite which might not be sufficient to support the aggregate input traffic. Hence some buffering on the ground terminal is required. Furthermore mapping to protocols such as Synchronous Digital Hierarchy (SDH) may be required. This module provides an interface (similar to ATM-AM) between the terrestrial network and the satellite ground-station.

4.1.4 Satellite Constellations

The following considerations must be taken into account in the design of the satellite constellation: the orbits in which the satellites will be placed and the total number of satellites which will be required to provide coverage to the chosen service area.

The total number of satellites required in the constellation depends on the type of orbits and the design method, together with the minimum elevation angle to the ground terminal which is required by the system. The minimum elevation angle will determine the probability of shadowing by buildings, trees and so on and consequently the level of the signal fade experienced. Naturally, the cost of launching the satellites and the capability to do so must be taken into account. It is more expensive to launch satellites into high orbits than low orbits and less payload mass can be launched into higher orbits. Furthermore the satellite payload architecture, discussed in Section 4.2.5, greatly affects the cost of the satellite and the inter-working complexity. Readers can refer to Appendix A for a more detailed discussion on different orbital types and constellation geometry.

4.1.4.1 Impact of Satellite Constellation on System Performance

The choice of a satellite constellation will be guided by the need for coverage, availability and the path delay and loss characteristics.

Coverage
A satellite system does not have to cover the whole Earth in order to connect the over-

whelming majority of networks in the world. There is not much demand for network inter-connection at present at the North and South Poles and in the oceans. If a satellite constella-tion design is found which can minimise satellite numbers as well as covering all land-masses, then this might have an advantage over other constellation designs. Satellite capabil-ities could be tailored to the particular area which is being served, although with a non-geostationary system this concept might be difficult to achieve.

Availability
From a dependability point of view, the satellite link should have the following properties:

- The proportion of time during which it is in a down state (i.e. unable to support a transaction) should be as low as possible;
- Once a connection has been established, it should have a low probability of being either terminated (because of insufficient data transfer performance) or prematurely released (due to the failure of a network component).

We define availability as the proportion of time during which the satellite link is able to support a connection. Conversely, unavailability of a satellite link is the proportion of time during which the satellite link is unable to support a connection (i.e. it is in the down state) [8].

The total availability of the satellite network (A_{total}) is dependent on the availability of the satellite ($A_{satellite}$), the availability of the satellite link ($A_{propagation}$) and the availability of the satellite resources ($A_{congestion}$).

$$A_{total} = A_{satellite} \times A_{propagation} \times A_{congestion} \tag{4.1}$$

Satellite availability results from the combination of two independent aspects: satellite reliability and operation and maintenance (O&M) strategies adopted in the space segment.

The space segment O&M policies implemented in a system, are tightly related to business decisions which are specific to any particular satellite system. As a result, the methodologies for carrying out comparative evaluations among different alternatives must be tailored on the functional characteristics and economic constraints of a specific system. It would be very difficult to identify a general approach.

The modern satellites, which have been designed to be used in the constellation network, have become more and more complex in order to cope with the new demanding system requirements for the provision of innovative telecommunication services. The adoption of payloads with on-board regeneration, switching capabilities and on-board processing has surely increased the flexibility of the satellite and allowed the advent of the satellite constel-lation systems. On the other hand, such flexibility needs to be carefully managed in order to maximise the level of performance which can be achieved with these sophisticated and expensive switching nodes.

One of the aspect to be monitored and managed is the occurrence of partial failures in the satellite network. The nature of the satellite makes the adoption of maintenance and replace-ment strategies, typical of the terrestrial environment, impossible. Internal redundancies are usually included in the satellite architecture due to the design phase for some of the critical devices, for which a failure would result in the impossibility of the whole satellite to perform the assigned function (i.e. catastrophic failure). The effects from the occurrence of a cata-strophic failure affecting a satellite of a constellation can be solved only by de-orbiting the failed satellite and by replacing it with a new one (either with a spare one already in-orbit, or with a new one to be launched from the Earth).

During the time frame required for the satellite replacement (depending on the system characteristics, it could vary from a couple of weeks up to some months), the QoS provided by the satellite network is affected. A careful design and management of a satellite constellation network should adequately take these parameters into account in order to be able to effectively guarantee to the end user a given level of service.

An additional problem comes from the possible occurrence of partial failures in the satellite. A partial failure consists of a failure affecting the capability of the satellite to properly perform one (or more) specific function, but does not prevent the satellite from performing most of its functions. In this case, the satellite is not likely to be de-orbited and, on the contrary, it will continue to work in a degraded way, until either a catastrophic failure or the natural end of the satellite cycle of life occurs.

Since the design lifetime of the modern satellite is increased with respect to the past, the probability of having several satellites in a constellation which are affected by different partial failures, could be not negligible. This mainly depends on the maintenance policy adopted in the constellation and normally results from some trade-off among QoS and cost/benefits). It therefore seems advisable to take care of the possible effects of these events by selecting adequate strategies, algorithms and methodologies for the management of the connections able to provide good performances, including in a degraded environment.

Path Delay

In a system which delivers multimedia services, the varying tolerance for delay means that different satellite orbits are suited to different types of service. Real-time services such as speech and video conferencing, which require low delay are more suitable to be delivered by a low orbit system; whereas non-real-time services such as ftp and fax are well-suited to higher-altitude satellites such as geostationary. It has to be noted that LEO systems have delay variation problems due to the fast movement of the satellite relative to the user. The service mix opens up the possibility of a hybrid system where certain services are delivered by the appropriate satellites.

Path Loss

The higher the orbit, the higher the path loss, since the free space loss is proportional to D^2, where D is the distance from the ground terminal to the satellite.

4.1.5 Satellite Payload Architecture

The satellite payload architecture with OBS and ISL will be very different from the conventional bent pipe structure. We will present the advantages and disadvantages of the OBP in this section and present a satellite payload reference functional architecture.

OBP satellites with high-gain multiple spot-beams and OBS capabilities have been considered as key elements of new-generation satellite communications systems. These satellites support small, cost-effective terminals and provide the required flexibility and increased utilisation of resources in a bursty multimedia traffic environment.

New possibilities arise with the future introduction of OBS and the use of new channel coding schemes. OBS transforms the pure reflector-type satellite into a switching node in the sky. This allows packets from several sources to be routed to several destinations, which is a meshed configuration and perhaps even more flexible than the optical node because of the inherent broadcasting possibility of a satellite. OBS will, in addition to switching, make error correction on-board the satellite, thus either reducing the BER or liberate capacity due to the possibility of discarding packets in the satellite which will not arrive at destination with the required QoS. In addition, judicious choices of FEC codes will make high quality transmission possible even with small antenna devices.

The advantages rendered by the use of OBP are summarised [9,10]:

• Regenerative Transponders

The advantage of the regenerative scheme is that the uplink and downlinks are now separated and can be designed independently of each other. With conventional satellites $(C/N)_U$ and $(C/N)_D$ are additive: whilst with regenerative transponders they are separated. This can be translated into an improved BER performance due to the reduced degradation. Regenerative transponders can withstand higher levels of interference for the same overall $(C/N)_T$.

• Multirate Communications

With OBP it is possible to change between low- and high-rate terminals on the satellite. This allows ground terminals operating at various rates to communicate with each other via a single hop. Transparent transponders would require rate conversion terrestrially and hence necessitate two hops. Multirate communications implies both multicarrier demodulators and baseband switches.

• Reduced Complexity Earth-Stations

Although employing an on-board switch function results in more complexity on-board the satellite, the following are the advantages of on-board switches.

• Lowering the ground station costs.
• Providing bandwidth on demand with half the delay.
• Improving interconnectivity.
• Offering added flexibility and improvement in link performance.

One of the most critical design issues for on-board processing satellites is the selection of an on-board baseband switching architecture. Four types of on-board switches are proposed:

• Circuit switch
• Fast-packet switch
• Hybrid switch
• Cell switch (ATM switch)

From an efficiency-of-bandwidth point of view, circuit-switching is advantageous under the condition that the major portion of the network traffic is circuit-switched. However, for bursty traffic, circuit-switching results in a lot of wasted capacity.

Fast-packet switching is an attractive option for a satellite network carrying both packet-switched traffic and circuit-switched traffic. This is particularly attractive for IP as the recent QoS framework by the Internet Engineering Task Force (IETF) makes IP routing an interesting alternative to ATM.

In some situations, a mixed-switch configuration, called hybrid switching and consisting of both circuit and packet switches, may provide an optimal on-board processor architecture. However, the distribution of circuit- and packet-switched traffic is unknown, which makes the implementation of such a switch a risk.

Finally, fixed-size fast-packet-switching, called cell-switching, is also an attractive solution for both circuit- and packet-switched traffic. Using statistical multiplexing of cells, it could achieve the highest bandwidth efficiency despite a relatively large header overhead.

The advantages and disadvantages of different on-board switching techniques are summarised in Table 4.1.

OBS introduces some issues for the design and analysis of the satellite architecture. Many considerations which were previously the concern only of the ground segments now shift to the space segment. The on-board processor allocates bandwidth on demand and performs statistical multiplexing. This essentially changes the nature of the satellite from a deterministic system to a stochastic system. In a stochastic system, the arriving traffic is random and statistical fluctuations may cause congestion where cell loss due to buffer overflow might occur. Thus, it is necessary to incorporate traffic and control mechanisms to regulate the input traffic.

4.1.6 Satellite Payload Reference Functional Architecture

The recent applications of the on-board processing concept has taken full advantage of the new possibility offered by fast digital signal processing technique and by advances in integrated high speed electronic technology. The overall weight, power consumption and cost of a satellite payload has been reduced considerably with enormous benefits in terms of overall cost and performance. By reducing the overall weight of the payload, more processing power has been allocated on-board. This has enabled a better integration with the ground network which can interface the satellite as a node able to access the same network resources with much more flexibility. Satellites designed with this innovative approach will therefore be provided with features able to adequately cope with the requirements of the future high speed networks.

The complexity of the Earth and space segment in the design of new satellite systems, has been reversed over the years, by increasing the complexity of the payload in order to limit costs and dimensions of the ground stations with the deriving economical benefits. Multi-beam antenna coverage is adopted which requires the introduction of base-band OBS (circuit or packet) necessary to ensure the full interconnectivity between different spot beams of the same satellite. The satellite contains all the functions necessary to route communication traffic through the network, including Earth-to-space, space-to-Earth and space-to-space connections. With this architecture, a signal received by a satellite may be transmitted directly back to Earth in the same or in a different beam or relayed by radio or optical inter-satellite links through other satellites from which it is then transmitted to Earth.

During COST253 Action, an attempt to identify a general reference functional architecture for the satellite payload was made. The aim of this was to identify a common baseline which could be used in the different modelling and simulation activities in progress and covering mainly access, routing, reliability and traffic aspects. The identification of such a general functional architecture requires the adoption of some assumptions summarised in the following [11]:

Table 4.1 Comparison of various switching techniques

Switching architecture	Circuit switching	Fast-packet switching	Hybrid switching	Cell switching (ATM switching)
Advantages	Efficient bandwidth utilisation for circuit switched traffic	Self-routing	Handles a much more diverse range of traffic	Self-Routing with a small VC/VP
	Efficient if network does not require frequent traffic reconfiguration	Does not require control memory for routing	Optimisation between circuit switching and packet switching	Does not require control memory for routing
	Easy to control congestion by limiting access into the network	Transmission without reconfiguring of the on-board switch connection	Lower complexity on-board than fast-packet switch	Transmission without reconfiguring on-board switch connection
		Easy to implement autonomous private networks	Can provide dedicated hardware for each traffic type	Easy to implement autonomous private networks
		Provides flexibility and efficient bandwidth utilisation for packet switched traffic		Provides flexibility and efficient bandwidth utilisation for all traffic sources
		Can accommodate circuit-switched traffic		Can accommodate circuit-switched traffic
				Compatible with ATM standards
				Fixed size packets

Disadvantages	Reconfiguration of Earth station time/frequency plans for each circuit set-up	For circuit switched traffic higher overheads than circuit switching due to packet headers	Can not maintain maximum flexibility for future services because the future distribution of satellite circuit and packet traffic is unknown	Speed comparable to fast-packet switching
	Fixed bandwidth assignment (not flexible)	Contention and congestion may occur	Waste of satellite resources in order to be designed to handle the full capacity of satellite traffic	For circuit switched traffic somewhat higher overheads than packet switching due to 5 byte ATM header
	Very inefficient bandwidth utilisation when supporting packet-switched traffic			Contention and congestion may occur
	Difficult to implement autonomous private networks			

- The communication payload is regenerative;
- It includes on-board processing and packet switching functions;
- Inter-satellite links are present among satellites of the same constellation;
- User access is provided by means of multi-spot coverage;
- Independent overlapping coverage using non-overlapping frequency bands and different spot contours could be present to accommodate ground terminals of different classes (i.e. small terminal vs. gateways);
- Coverage could be provided either by moving spots on ground (fixed pattern generated by the on-board antenna) or by fixed spots on ground (adoption of the beam steering technique).

The functional architecture of the payload could be represented as shown in Figure 4.4. The actual architecture of a satellite (number of beams, RX/TX chains, presence of different types of coverage, internal architecture of the various subsystems, ... etc.) varies from system to system depending on system design choices and requirements. Usually, details about the actual architecture are confidential and it is difficult to access them. For this reason, the modelling and simulation activity being carried out in the COST253 Action, plans to define a general methodology that should be detailed and customised when applied to a real system and when the relevant technical details become available.

The functional architecture of the base-band OBP matches the structure of the IF and RF components in Figure 4.4. We assume (which could be unrealistic for certain satellites) that each link is assigned a dedicated, independent link processor. The required processing capability is embedded in both the base-band uplink and downlink processors and the base-band switch.

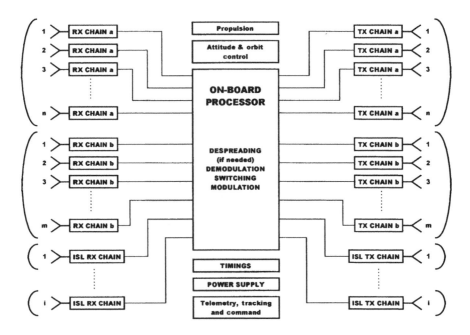

Figure 4.4   Functional architecture of a general payload

4.1.7 ISLs

The inclusion of a satellite on-board switch with some ATM functionality was considered for the satellite architecture within COST253. These switches would route packets (or ATM cells) using the information in the header. Options on-board the satellite for routing, include routing via individual spotbeams to ground stations, or via ISLs to other satellites which will further route the packets. Advantages and disadvantages of ISLs are summarised below.
Advantages of ISLs:

- Calls may be grounded at the optimal ground station through another satellite for call termination – reducing the length of the terrestrial 'tail' required.
- A reduction in ground-based control may be achieved with on-board baseband switching – reducing delay (autonomous operation).
- Increased global coverage – oceans and areas without Earth stations.
- Single network control centre and Earth station.

    Disadvantages of ISLs:

- The complexity and cost of the satellites will be increased.
- Power available for the satellite/user link may be reduced.
- Handover between satellites due to inter-satellite dynamics will have to be incorporated.
- Replenishment strategy.
- Frequency co-ordination.
- Cross-link dimensioning.

    Despite these disadvantages, the advantage of routing traffic in the sky independently of the ground infrastructure, makes the use of ISLs an attractive solution.

4.2 Resource Control

4.2.1 Resource Allocation

COST253 has supported follow-on studies to resource allocation schemes that had been investigated by two teams participating in COST226 and subsequently in COST253, namely CNUCE (Italy) and ENST (France). Within COST226, the considered resource allocation schemes had been designed for efficient interconnection of LAN by satellite links. The traffic delivered by such LANs was considered an aggregate traffic and two main components have been identified: real-time traffic (telephony, video) and non-real-time traffic (computer data exchange). Real-time traffic is generated at a nearly constant bit rate and is often referred to as stream traffic. Non-real-time traffic consists typically of data bursts with silence in between and is called bursty traffic. Within bursty traffic, one identifies two classes: bulk traffic and interactive traffic, depending on the respective value of burstiness, (typically 20–900 for bulk traffic and, 4000–5000 for interactive traffic) [12].
    A different approach has been considered within COST253, which is relevant to the user access mode. In this mode, satellite networks provide broadband access to end users by means of small Earth stations (often quoted as Very Small Aperture Terminals (VSATs)) and asymmetric links, as illustrated in Figure 4.5. The user Earth station sends requests to a gateway through a small capacity link (typically a few hundred kb/s) and receives delivery of requested data from a gateway through a larger capacity link (typically tens of Mb/s). The

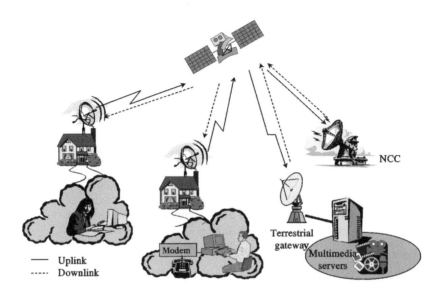

Figure 4.5   Configuration of user access mode

high capacity link from the gateway to the user Earth station is called the forward link, while the opposite low capacity link is the return link. This mode is envisaged for providing universal personal services and interactive multimedia applications to residential as well as professional users by connecting them to remote terrestrial information sites. Network control management and resource allocations functions are centralised, either in an Earth station acting as the network control centre (NCC) or on-board the satellite.

This mode differs from the LAN interconnection in the sense that traffic multiplexing coming from different terminals connected to user Earth stations in different locations, must be done at the air interface and not prior to transmission, as these terminals are not part of a terrestrial LAN. More specifically, a terminal connected to a given user Earth station generates traffic according to the application running on that terminal. The multiplexing between different traffic types, generated by different terminals, is then performed over the radio link under the control of the NCC and there is no traffic aggregation at the user Earth station, as in the case in LAN interconnection. Therefore an efficient resource allocation scheme is necessary to share the uplink bandwidth resource between different traffic types.

A novel resource allocation scheme, named Double Moveable Boundary Strategy (DBMS) has been investigated. DBMS is a dynamically controlled boundary policy which adapts the allocation decision to the variable network loading conditions and supports both connection-oriented and connection-less services. DBMS is based on a Time Division Multiple Access (TDMA) scheme, where the time frame is divided into three parts as illustrated in Figure 4.6. The NF time slots in the frame consist of C1 slots reserved for Constant Bit Rate (CBR) traffic, D slots reserved for bursty traffic and the remaining number of slots within the Common Resource Pool (CRP).

A terminal running a particular application transmits a call set-up request to the frame resource allocator located in the NCC or on-board the satellite, or a burst reservation request depending on the type of traffic generated by the application. This request is

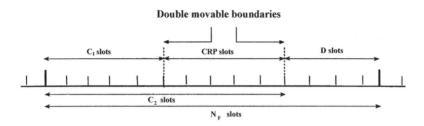

Figure 4.6   DBMS frame organisation

transmitted either on a separate signalling channel or on the uplink traffic channel, using signalling mini-slots.

The priority of each traffic category in each sub-frame is defined as follows:

- Sub-frame with C1 slots (isochronous traffic oriented): highest priority for CBR traffic, then bursty traffic and finally best-effort traffic.
- Sub-frame with D slots (asynchronous traffic oriented): highest priority for bursty traffic, then best-effort traffic (denied access to CBR traffic).
- CRP: relative priority of both isochronous and asynchronous traffic defined according to the DMBS algorithm.

The Connection Admission Control (CAC) and the slot allocation decisions are taken at the beginning of each control period, which consists of single or multiple TDMA frames. The frame resource allocator, located in the NCC or on-board the satellite, first allocates slots to waiting CBR traffic, according to waiting reservation requests. If time slots are available in the CBR sub-frame, the CBR traffic will be granted a slot throughout the call duration. If there are still requests waiting to be satisfied after all slots in the CBR sub-frame are occupied, CRP time slots will be allocated to this CBR traffic, under the condition that the reservation request queue is less than a pre-specified threshold.

Interactive applications occasionally generate a short burst of data and send resource reservation requests for the duration of the burst. These requests are queued until time slots are available on the frame. Bursts are allocated D slots from the sub-frame and after the exhaustion of the D slots, form the CRP and the CBR slots. Whenever the bursty data requests are all satisfied, the best effort traffic requests are taken into account and the remaining time slots, whichever sub-frame they belong to, are allocated to the best effort traffic.

Once the frame allocation plan is established, it is broadcast to all the active Earth stations over the downlink.

The DBMS has been extensively evaluated through analytical modelling and simulation. The detailed results are available from [13–20]. It has been shown that the data queue threshold value is an important parameter that can be dynamically controlled to adapt the allocation decision to varying traffic. Simple formulas have been proposed to adjust this threshold as a function of the number of time slots allocated to the CBR traffic in the CRP. The dynamic threshold policy reduces the extra delay to which data is subjected under the DMBS scheme with little impact on the CBR.

4.2.2 Call Set-Up and Routing

4.2.2.1 Routing Strategies

In deriving routing strategies, three main issues need to be considered.

The Location of Routing Control Functions

The routing control functions can either be located in a centralised, a decentralised or a distributed manner.

The centralised approach performs route computation in a single node of an entire network. The main disadvantages are that this approach is prone to a single point of failure and it needs a powerful dedicated route server. Furthermore, signalling traffic (the route requests) is concentrated in the vicinity of the route server. Finally, each node using the route server must know the default route to the server.

The distributed approach does not require powerful nodes since each node performs only local route computation. Signalling is part of the distributed computation. However, distributed routing algorithms suffer from transient or permanent loops in the routes. These loops are caused by a loose synchronisation among the nodes contributing to the route computation. In most cases, this can be avoided by using loop avoidance mechanisms, although these may slow down the convergence rate of the algorithm.

The decentralised approach enables route computation at the node closest to the user requesting a route. It does not suffer from the drawbacks of distributed algorithm. Nevertheless, they require more powerful nodes in comparison to the distributed algorithm, although less than that of the centralised approach. If adaptive routing is considered, a signalling protocol is also required to distribute and receive information about the other nodes.

Trade-off Between Routing Efficiency and Resource Utilisation Efficiency

Routing algorithms can be either static or dynamic. Static algorithms involve the computation of routes in advance. They do not update the routes as a result of changes in the network connectivity or congestion level. The main advantage of static routing is that it reduces the user waiting time for route connection. However, the main drawback is its inefficiency in resource utilisation since the pre-computed route will be dedicated to the user for as long as the route is connected regardless of the route usage. Furthermore, route pre-computation is not always tractable for large scale networks.

Adaptive routing algorithms use information obtained from the network in order to update the routes whenever it is needed. This information advertise changes of a link status or resource depletion in a node. It is also possible to have pseudo-adaptive routing algorithms which use predictive models to adapt the routes either to connectivity changes or traffic load.

Adaptive routing can be further classified as isolated and non-isolated algorithms. Isolated algorithms only use local information, i.e. available from the node performing the route computation. Non-isolated algorithms use information gathered from adjacent nodes or from the whole network.

The Choice of Path Computation Mechanisms Based on Path Metrics and Path Selection Policies

The essence of routing is to provide a path from a source to a destination taking into account the user requirements. There are different ways to deal with these requirements which can be categorised as legacy routing, Type of Service (ToS) routing and QoS routing.

Legacy routing involves computing a route that minimises a given metric such as the number of hops or the end-to-end delay. This metric is 'hard-write' in the algorithm and cannot be changed.

ToS routing is the extension of legacy routing to multiple metrics. When requesting a route, the user specifies the metric according to which route will be optimised. Both legacy and ToS routing fall in the category of best effort routing where the network provides the best route available. However, best effort routing may not produce the desired effect for today's telecommunication services. Services such as teleconferencing are more interested in low jitter than short delays.

The last category is QoS routing, which enables the user to express precisely their needs. Required end-to-end delay may, for example, be specified as an interval of acceptable values.

This shift of paradigms from legacy routing to QoS routing requires the route computations to be reconsidered. While shortest path algorithms, such as Dijkstra or Bellman Ford, are well-suited to best effort routing, it does not apply to QoS routing. Rather than a shortest path, a 'best fitting path' must be found. Furthermore, the fitting function has to take many constraints into account, whilst the shortest path is restricted to a single metric. Algorithms supporting QoS routing are known in the operation research domain as multi-criteria optimisation procedures. Problems in this area are theoretically NP-complete and makes use of heuristics to make the computation complexity tractable.

As far as QoS routing is concerned, on-demand routes are often used since it is difficult to compute every possible user request. Some implementation of distributed algorithms, namely distance vector, are not well-adapted to on-demand routing, therefore restricting their use for QoS routing.

4.2.2.2 Issues in the Implementation of Routing Strategies in Non-Geostationary Networks

Traditionally satellites have been extensively used for broadcast services and telephone trunk links, both taking advantage of satellites positioned in GEO. Recently the telecommunication industry faced a challenge to provide a variety of new, broadband multimedia services for users equipped with fixed and mobile terminals. Requirements for higher capacity and lower propagation delay made non-geostationary satellite constellations appealing, especially with advances in technology which enabled the implementation of ISLs. Many satellite communication systems have been proposed in the last few years, both for the provision of mobile telephony and internet-in-the sky. Several of these proposals already incorporate ISLs suitable for traffic interconnection in the satellite segment of the network.

However, non-geostationary satellite systems with ISLs still lack efficient routing algorithms, adaptive to inherent dynamics of topology and traffic load. Current solutions are mainly reusing algorithms developed and optimised for the use in terrestrial networks with static topology, thus having only limited capability to grasp the characteristics of non-geostationary satellite system. In addition, dynamic topology of satellite networks and variations in traffic load in satellite coverage areas due to the motion of satellite in their orbits, pose stringent requirements to routing algorithms. Therefore, routing algorithms incorporating a traffic flow model, which take into account the geographic distribution of traffic sources and time variation of traffic intensity and of the ISL model, have to be derived in non-geostationary satellite communication systems.

Various simulations have been carried out to study the performance of different routing algorithms in MEO and LEO satellite networks. In general, the simulation models consist of the following components:

- The satellite system dynamics components which describe the satellite constellation characteristics;
- The traffic simulation component which considers the geographical distribution of traffic sources and the daily variation of their traffic intensity as well as the traffic source generation;
- The ISL network component which studies the performance of simulated routing algorithm under various network conditions.

Under COST253, three case studies have been performed to investigate the routing strategies for non-geostationary satellite networks. The details of the studies and their simulation results are reported in the following section.

4.2.2.3 Performance Evaluation of Routing Algorithms in Non-Geostationary Networks

Case Study I
Satellite System Dynamics
In this study, both MEO (Odyssey-like) and LEO (Celestri-like) satellite constellations are investigated. The parameters of the selected satellite systems are summarised in Table 4.2.

The satellite dynamic topology is considered through a set of time-discrete snapshots of satellite positions; a technique which is frequently used [30–32]. The position of satellites at a given snapshot is calculated in respect to the longitude and latitude of the sub-satellite point on the surface of the Earth using equations in [33]. The time step chosen for the snapshots has to guarantee a reasonably smooth sequence of snapshots. In the case of the Odyssey-like MEO constellation, a 5-min time step is selected, which enable satellites to pass approximately 5% of their footprint diameter. For the Celestri-like LEO constellation, a 1-min step enabling satellites to pass 7% of their footprint diameter has been selected. Figures 4.7 and 4.8 show the snapshot of the Odyssey-like and Celestri-like constellation with ISLs respectively.

Table 4.2 Parameters of the selected satellite systems

	Odyssey-like constellation	Celestri-like constellation
Orbit altitude	10354 km	1400 km
Orbit period	6 h	114 min
Orbit inclination β	55°	48°
Number of satellites	12	63
Number of orbits	3	7
Number of ISLs	2 intra-orbit + 2 inter-orbit	2 intra-orbit + 2 inter-orbit
Minimum elevation angle	20°	16°

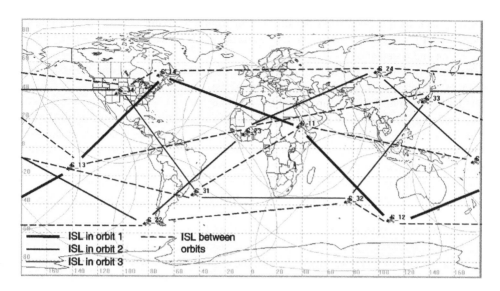

Figure 4.7   Snapshot of Odyssey-like constellation with ISLs

Dynamic Traffic Distribution

Besides the dynamic topology, non-geostationary satellite systems are subject to the dynamic variation of traffic load with both changing satellite location and time of the day. This dynamic is also deterministic and can be computed in advance before the start of the simulation.

The traffic intensity greatly depends on the geographical location on the Earth. Although designers of different satellite communication systems target customers with different profiles, they generally agree on several parameters that have the highest impact on traffic

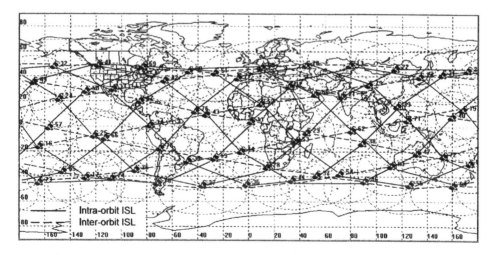

Figure 4.8   Snapshot of Celestri-like constellation with ISLs

intensity, such as the population density and distribution, living habits, Gross Domestic Product (GDP), industrial development, existing telecommunication infrastructure, service penetration and acceptance level, etc.

The traffic flow model has to take into account three phenomena causing the traffic flow dynamics; all being influenced by the selection of the service to be provided.

(1) Distribution of Traffic Sources and Destinations

For studying the inherent characteristics of the selected satellite constellation, the following points must be considered: (i) uniform distribution of traffic sources and destinations all over the surface of the Earth, or (ii) a landmass model with uniform distribution of sources/destinations over the continents and major islands. For more realistic conditions, however, a customised model of a non-uniform distribution has to be designed, taking into account the real geographic distribution of sources and destinations for the provision of the particular service.

(2) Traffic Flow Distribution Between Sources and Destinations

The traffic flows between sources and destinations are defined according to the selected source–destination flow table, which should reflect the flow characteristics for the selected type of service. Different tables have to be defined for different types of services. The source–destination flow table should provide source–destination pairs at least on the level of the Earth's six continental regions, as defined in [30], if not between countries/territories. Table 4.3 gives an example for source–destination flow table typical for telephony service.

(3) Daily Variation of Traffic Intensity

The daily variation of traffic intensity in non-geostationary satellite systems is caused by the activity variation of the user due to local time of the day, which in turn depends on the geographical time zone of user location. Thus, the daily variation can be captured by an appropriate profile of daily user activity, as shown in Figure 4.9, together with a simplified geographical time zone model, which ignores country borders and increments one local hour for each 15° longitude.

Based on these models a geographical distribution of relative traffic source intensity can be calculated for any location on the surface of the Earth and mapped to the appropriate source satellite as the sum of traffic contributions within its coverage area. After the definition of destination region, the traffic is proportionally divided among the satellites according to their coverage in that region. Such an approach is suitable especially for the case of studying only ISL network segment, since it avoids problems pertaining to the satellite access and the generation of traffic on the level of individual user.

Table 4.3 Percentage of total traffic flow between source and destination region

	North America	South America	Europe	Africa	Asia	Oceania
North America	85	3	4	2	4	2
South America	7	81	7	2	2	1
Europe	4	3	85	3	4	1
Africa	5	2	7	81	4	1
Asia	5	1	5	2	83	4
Oceania	5	1	2	1	7	84

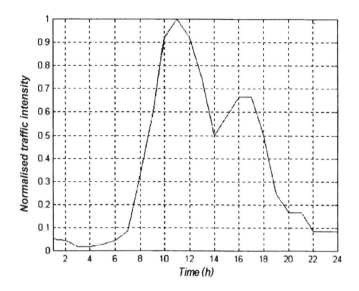

Figure 4.9   Daily profile of user activity

Traffic flow dynamics, affected by rotation of the Earth and the user activity profile, typically has a period equal to one Earth day. Due to the periodic change of the topology and traffic flow dynamics, satellite systems also exhibit periodicity. Thus, in order to take into account all possible combinations of variable topology and traffic flows, it is sufficient to simulate only one system period – defined in [30] as the smallest common integer multiple of the orbit period and one Earth day (24 h).

For the traffic source our simulator enables us to use an optional model, but in the first phase only the Poisson generation process is implemented, although it does not reflect the conditions in real broadband networks. In addition to randomly generated packets with the Poisson process, reply packets are also generated in destination node, thus having a mechanism to control traffic asymmetry.

ISL Network Model   The ISL network simulator should not be restricted to a certain communication protocol or connection mode. On the contrary, it has to consider the specific characteristics of a packet or cell switching communication system, especially those having considerable impact on performance of different routing algorithms. The ISL network simulator has been built on the packet level, which increases the system complexity and computational effort. It also allows studying adaptive routing algorithms considering actual status of the network, such as delay or congestion, recognised by a node from the incoming packets.

The ISL network simulator has many simulation parameters to study the performance of simulated routing algorithm under various network conditions. Some of them are: mean and maximum packet length; capacity of intersatellite link; total network traffic load; traffic symmetry factor; routing matrix update period; time between snapshots; weight factors for the impact of traffic load and delay to link cost.

Our main goal during development of this simulator was to get an insight in the simulated system and to learn more about the problems pertaining to traffic routing in highly dynamic non-geostationary satellite systems. The simulator also enables us to study the effect of using different assumptions regarding: geographical distribution of traffic sources; daily profile curve; source–destination traffic flow tables; traffic source model; traffic asymmetry model; parameters and weight factors for link cost calculation.

The ISL network module, which is the core of the ISL network simulator, consists of a traffic source generator and of communication functions of the network. In this module data packets are actually generated considering the relative traffic intensity experienced by a particular satellite. They are routed towards the destination node taking into account the selected routing algorithm. Finally, before these packets are terminated, the statistical data is collected for routing table updates and for postprocessing of simulation results.

The ISL network simulator allows implementation of an arbitrary traffic source generator. Initially only a generator based on the well-known Poisson generation process has been implemented, but in the later stage the use of a more realistic traffic generator is envisaged. Besides randomly generated packets with the selected traffic source generator, it is hoped to generate reply packets in the destination node, thus having a mechanism to shape the traffic asymmetry. The level of traffic asymmetry is controlled with a simulation parameter reply factor (RF), which actually defines the amount of the replying traffic with respect to the incoming traffic.

Regarding the communication functions, only a limited functionality of the first four OSI layers is implemented in the ISL network simulator, as described in the following.

In the transport layer, the source satellite generates packets with all the necessary data for routing and later for the analysis. In the destination satellite the same layer is responsible for terminating the incoming packets and for generating reply packets according to the selected symmetry shaping.

Routing tables are computed with the selected routing algorithm in networking layers. These tables are used for routing of packets towards the next node on the shortest path to their destination. As a starting point a centralised version of the Dijkstra shortest path algorithm has been implemented.

In data link layer packets are actually sent to and received from ISL. Prior to being sent to the appropriate intersatellite link, packets are put in a FIFO queue which enables a distinction of different packet priorities. The data link layer also gathers information about the link cost, required for calculation of the routing table in the networking layer. A link cost metric is composed of traffic load and the propagation delay on the link ($D_{propagation}$). Their relative impact on the Link Cost (LC) is linearly regulated with a Traffic Weight Factor (TWF) and a Propagation Delay Weight Factor (PDWF), respectively, as shown in Eq. (4.1). In fact, the impact of the traffic load on link cost is considered via the delay in the satellite output queue ($D_{queuing}$).

$$LC = PDWF \cdot D_{Propagation} + TWF \cdot D_{Queuing} \tag{4.2}$$

On the physical layer, only the length of ISLs is taken into account, expressed as the propagation delay between pertaining satellites.

Simulation Results

Uniform Traffic   The simulator can provide various results, such as average delay experienced by packets in the network; average number of hops between origination and destination

Table 4.4 Simulation parameters for uniform traffic

	Celestri-like constellation
Reply factor	{0, 1}
Traffic weight factor	{0, 1, 10}
Propagation delay weight factor	1
Mean packet length	1000 bit
Maximum packet length	10000 bit
Link capacity	100000 bit/s
Total network source traffic	100/200 packets/s
Simulation duration	6840 s
Snapshot duration	60 s
Number of snapshots	114
Refresh cost matrix window size	30 s
Refresh routing rate	30 s

satellite node; total number of sent and received packets and utilisation of certain intersatellite links.

Table 4.4 summarises the simulation parameters used in the ISL network model for uniform traffic simulation.

For uniform traffic distribution, only results for Celestri-like constellation are presented. The following assumptions have been made:

- Uniform distribution of traffic sources and destinations over the surface of the Earth.
- A uniform source–destination traffic flow table.
- A constant daily profile curve.
- Poisson traffic source generator.
- Symmetry shaping for an unidirectional traffic flow (RF = 0) and a symmetric bi-directional traffic flow (RF = 1).
- Dijkstra shortest path algorithm for central calculation of routing tables.
- The duration of the simulation run equal to one orbit period.

First, average packet delay and average number of hops in the Celestri-like network are presented for different values of RF and TWF in Figures 4.10 and 4.11, respectively. In order to make a fair comparison between the results for the simulation runs with (RF = 1) and without (RF = 0) replying packets, twice as many packets have been generated in case of RF equal to zero, thus guaranteeing the similar traffic load in the network except for the symmetry. The results show that the traffic symmetry under assumed conditions does not have any significant influence on either average delay or average number of hops. The increase of relative impact of traffic load on the link cost for the calculation of routing tables, however, has a slightly more noticeable effect to results for average delay and average number of hops. The differences for different values of TWF factor are still negligible, which is a consequence of the high connectivity of the ISL network, providing several routes between a pair of nodes with the same number of hops and thus similar propagation delay.

The impact of TWF to the link cost assuming symmetric or asymmetric traffic load can be

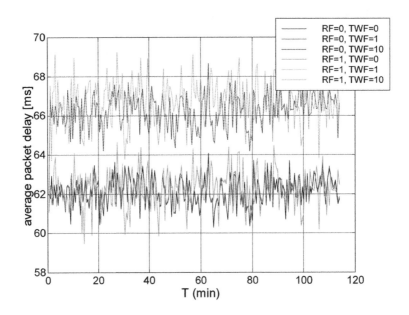

Figure 4.10 Average packet delay in the Celestri-like network for asymmetric (RF = 0) and symmetric (RF = 1) traffic load

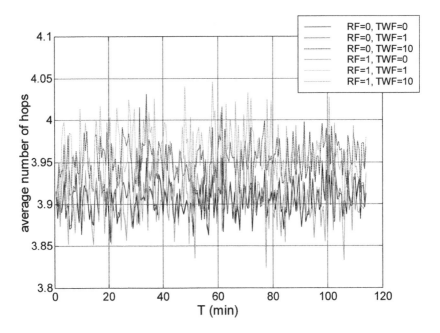

Figure 4.11 Average number of hops in the Celestri-like network for asymmetric (RF = 0) and symmetric (RF = 1) traffic load

Table 4.5 Queuing delay statistics for all ISLs

		Mean (ms)	SD (ms)	Minimum (ms)	Maximum (ms)
RF = 0	TWF = 0	0.431	0.445	0	5.349
	TWF = 1	0.431	0.446	0	5.349
	TWF = 10	0.4854	0.523	0	18.22
RF = 1	TWF = 0	0.457	0.464	0	5.526
	TWF = 1	0.458	0.466	0	5.526
	TWF = 10	0.484	0.556	0	17.0

estimated using Eq. (4.1). Typical values are taken into account for propagation delay in the selected constellation between 10.3 and 19.7 ms and the statistics for the queuing delay, which are summarised in Table 4.5.

Apparently, the traffic load in case of TWF = 10 can have a higher impact on the calculation of routing tables than propagation delay, thus enabling routing via longer but less burdened links. This is confirmed by the results in Figure 4.12 showing the mean delay of packets received in one orbit period by each of 63 satellites in the network. Packets received by a particular satellite in case of TWF = 10 exhibit longer delays than in the case of TWF equal 0 or 1 regardless of traffic symmetry.

Furthermore, the variation of queuing delay as reported in Table 4.5 indicates that by using

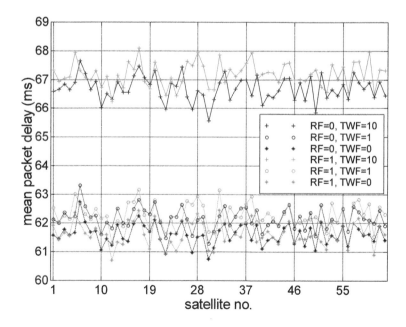

Figure 4.12 Mean delay of received packets on satellites in one orbit period for asymmetric (RF = 0) and symmetric (RF = 1) traffic load

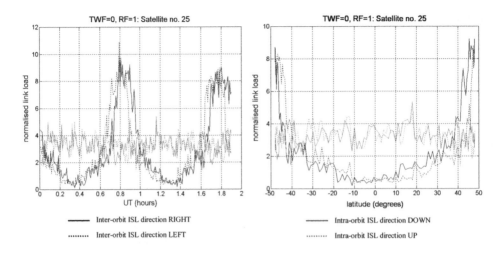

Figure 4.13   Satellite with the highest normalised link load for TWF = 0 and RF = 1

fixed weight factors the ratio between traffic load and the propagation delay in the link cost metric is changing during the simulation run. In order to keep this ratio within certain limits, weight factors should be adaptive to topology and traffic load variations.

Figures 4.13–4.15 illustrate satellites with the highest normalised link load for symmetric traffic load (RF = 1) and for different values of TWF in dependence of time and of the latitude of the subsatellite point. Link load is calculated from the throughput of the ISL in certain time-step, normalised with the link capacity. Similar results have been also obtained for asymmetric traffic load (RF = 0). Table 4.6 summarises results for the highest normalised link load identifying the ISL, the time of the occurrence of the peak and the location of the corresponding satellite.

The results for TWF = 0 in Figure 4.13 show the fluctuation of normalised link load due to

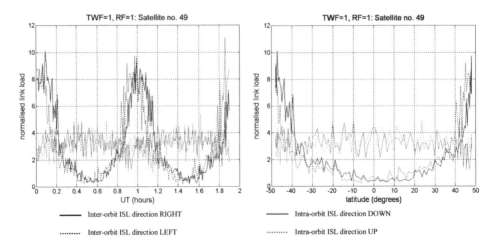

Figure 4.14   Satellite with the highest normalised link load for TWF = 1 and RF = 1

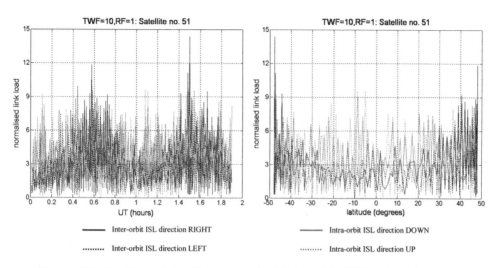

Figure 4.15   Satellite with the highest normalised link load for TWF = 10 and RF = 1

updating of routing tables considering only propagation delay. A significantly higher fluctuation is experienced in case of a high impact of traffic load on the link cost. Results in Figure 4.15 for TWF = 10 show unstable network conditions, since routing tables are changing significantly after every update interval equalling 30 s. Figures 4.13–4.15 also indicate that regardless of simulation parameters the highest normalised link load is always achieved on the interorbit links (left or right ISLs). The periodicity, demonstrated for interorbit links in case of TWF = 0 and TWF = 1, indicates the impact of satellite location on routing tables. Results for low impact on link cost of traffic load relative to propagation delay (TWF equal 0 and 1) clearly indicate that most of the traffic tends to be routed between orbits at higher latitudes, where the distance between the neighbouring satellites decreases. Results for TWF = 10, on the other hand, show higher link utilisation also for intraorbit links and at the latitudes around equator, due to lower impact of propagation delay on the calculation of routing tables.

Furthermore, as shown in Figures 4.13–4.15, in case of uniform distribution of traffic sources and uniform source–destination traffic flow tables, the link utilisation is periodic with half of the orbit period, or with 180 degree latitude.

Table 4.6   The highest normalised link load for different values of TWF in case of symmetric traffic load (RF = 1)

TWF	The highest normalised link load			
	Value	ISL	Universal time	Latitude
0	10.887	Inter-orbit left (Sat 25 to Sat 17)	0 h 48 min	47.93°
1	11.054	Inter-orbit left (Sat 49 to Sat 41)	1 h 52 min	−46.07°
10	14.383	Inter-orbit right (Sat 51 to Sat 59)	1 h 30 min	−47.85°

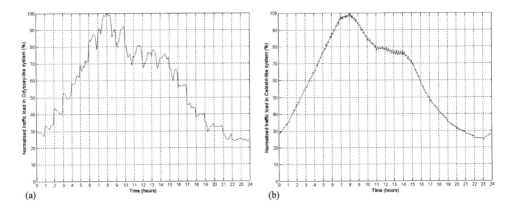

Figure 4.16 Normalised total traffic load for (a) Odyssey-like constellation and (b) Celestri-like constellation

Non-Uniform Traffic   In case of non-uniform traffic load conditions, the performance of the routing algorithm has been measured in terms of average delay experienced by the packets in the network; the average number of hops between origination and destination satellite node; the total number of sent and received packets; and utilisation of certain inter-satellite links. Some representative results are given for both Celestri-like and Odyssey-like constellations. Traffic loads in the period from 06:00 hours to 12:00 hours for Odyssey-like constellation as shown in Figure 4.16a and in the period from minute 0 to minute 206 for Celestri-like constellation as shown in Figure 4.16b have been used. In the latter, this is the time elapsed before two different satellites pass the same point on the Earth [34]. The simulation parameters are summarised in Table 4.7.

Figures 4.17 and 4.18 show respectively the average delays in the Odyssey-like network and the Celestri-like network using different values of reply factor. As can be seen from the results, the higher the reply factor, the higher the average delay in the network. The difference

Table 4.7 Simulation parameters

	Odyssey-like constellation	Celestri-like constellation
Reply factor	{0,1,2,4}	{1,2}
Traffic weight factor	1	1
Propagation delay weight factor	1	1
Mean packet length	1000 bits	1000 bits
Maximum packet length	10,000 bits	10,000 bits
Link capacity	100,000 bits	100,000 bits
Total network source traffic	50 packets/s	90 packets/s
Simulation duration	21,600 s	12,360 s
Snapshot duration	300 s	60 s
Number of snapshots	72	206
Refresh cost matrix window size	60 s	30 s
Refresh routing rate	60 s	30 s

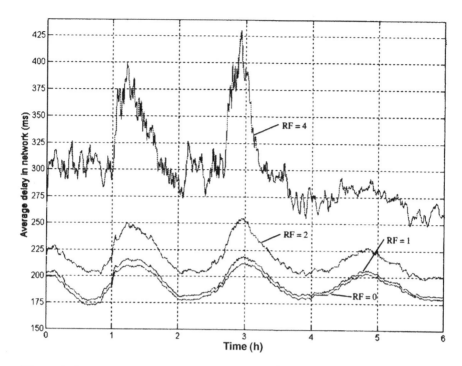

Figure 4.17   Average delay in Odyssey-like network for different values of reply factor

between different curves in Figure 4.17 representing simulation run for reply factors equal to 0 and 1 are almost the same. This is due to the fact that when reply factor is equal to 1, separate link is used for each traffic direction and both links are symmetrically loaded. For a higher reply factor, more return packets are generated and their packet lengths are longer than incoming ones, thus the time taken by the return packets during traversing the link is longer. As a result, certain links get more loaded and the average delay in the network increases.

   Another factor that affects the average delay is the propagation delay. Although the packets experience longer delay when a higher reply factor is applied, the average number of hops experienced by the packets in traversing from the source node to the destination node can hardly be distinguished, as shown in Figures 4.19 and 4.20. This implies that propagation delay has a much bigger impact in defining the link costs and hence the definition of the routing table since the packets use similar routes regardless of the reply factor.

Case Study II
Satellite System Dynamics

In this study, a Celestri-like satellite constellation is investigated. The parameters of the selected satellite systems are quite similar to those in Case Study I except for the number of ISLs in inter-orbits and are summarised in Table 4.8.

Dynamic Traffic Distribution and Traffic Source Generation

The approach adopted to take into account the dynamic traffic distribution is quite similar to that for Case Study I. In this study, 50 Earth stations are assumed to be distributed among the

Table 4.8   Satellite constellation parameters

	Celestri-like Constellation
Orbit altitude	1400 km
Orbit period	114 min
Orbit inclination β	48°
Number of satellites	63
Number of orbits	7
Number of ISLs	2 intra-orbit + 4 inter-orbit

six regions of the globe: Africa, Asia, Europe, North America, Oceanic and South America. Tables 4.9 and 4.10 show the distribution of Earth stations and the traffic level at each Earth station in terms of packets/s.

Furthermore, the traffic level at each Earth station is adjusted according to the daily variations of traffic intensity which is similar that shown in Figure 4.9.

Instead of adopting the Poisson model alone to generate the traffic source as it is in the previous case study, both the Poisson model and the self-similar model are used for traffic source generation at each station.

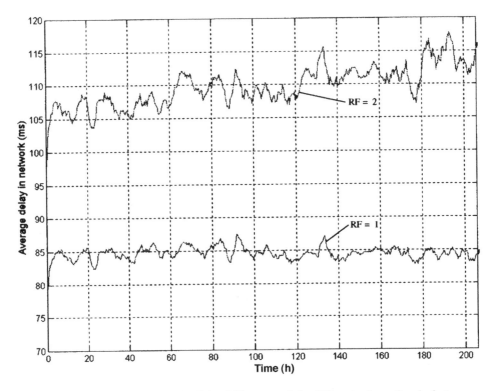

Figure 4.18   Average delay in Celestri-like network for different values of reply factor

The self-similar process is a mathematical model for the superposition of an infinity of on–off sources with Pareto distribution such that the burst length distribution of the random packets is of Pareto. Assuming that the sources produce packets with a constant rate R. If ω_s is the sth instant of the beginning of packet generation with the constant rate R and τ_s is an independent random variable denoting the packet generation interval starting at the s-th instant, the probability distribution function of τ_s is given by:

$$P\{\tau_s \leq t\} = \left[\frac{\beta}{t+\beta}\right]^\alpha \tag{4.3}$$

where $\beta > 0$ and $1 < \alpha < 2$. Denote the random variable ξ_t as the number of time moments, ω_s, such that $\omega_s = t$ and that it forms a Poisson process with intensity λ and they are independent of τ_s. Let the random variable Y_t denote the total rate of packet generation in the aggregate traffic at time t, the mean rate of packet generation at time t is given by:

$$E\{Y_t\} = R \times \lambda \times \alpha_t \tag{4.4}$$

and the variance is given by:

$$E\{(Y_t - E\{Y_t\})^2\} = R^2 \times \lambda \times \alpha_t \tag{4.5}$$

where α_t is the mean of the Pareto Distribution.

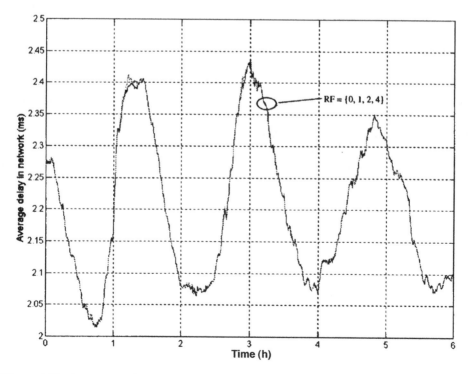

Figure 4.19   Average number of hops in Odyssey-like network for different values of reply factor

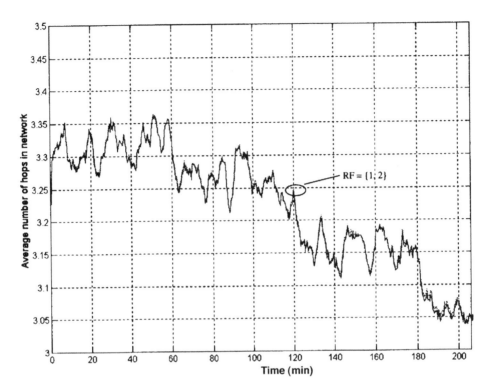

Figure 4.20 Average number of hops in Celestri-like network for different values of reply factor

ISL Network Model

In this case study, a Celestri-like system with six permanent ISLs is investigated. These ISLs lead to a topology with stable connections but time varying connection costs. In the study, ATM is used as the transport scheme. The Virtual Path (VP) is considered as the logical link between two adjacent ATM nodes, or satellites in this case. A VP is constituted by Virtual Circuits (VCs) between different user connections. A route between a pair of satellites (start and end satellite) may consist of subsequent ISL's and is modelled as a Virtual Path Connection (VPC).

Three routing algorithms have been investigated: the Dijkstra Algorithm [35]; the adaptive Dijkstra Algorithm [36]; and the Flow Deviation (FD) algorithm [35].

Dijkstra Algorithm   The cost function for Dijkstra Algorithm incorporates the transmission delay in a path and the time delay in each satellite hop. Denoting the transmission delay for the i-th satellite as D_i and the time delay for each satellite hop as H, the cost function C is given by:

$$C = \sum_{i=1}^{n} D_i + n \times H \tag{4.6}$$

where H is set to 10 ms and n is the number of satellites included in a path. The Dijkstra

Table 4.9 Location and magnitude of the Earth stations

Source number	Traffic Levels									
	1	2	3	4	5	6	7	8	9	10
Africa	0	0	2	0	0	0	0	0	0	0
Asia	0	6	3	0	1	2	1	1	1	1
Europe	0	6	1	2	1	0	0	3	0	0
North America	0	3	4	0	0	1	2	0	2	1
Oceania	1	0	1	0	0	0	0	0	0	0
South America	0	0	0	3	1	0	0	0	0	0

Algorithm is performed every 10 s to ensure any variation in topology is captured. The routing scheme is integrated with the minimisation of the switching between paths connecting a pair of satellites, that can be caused by the time variant effect of D, as proposed in [31].

Adaptive Dijsktra Algorithm   In order to find the k-shortest paths for an end-to-end connection between a pair of satellites, two basic rules are imposed in the adaptive Dijsktra Algorithm:

1. If the path chosen in the last time interval is available for the present one, it should be maintained as long as it's cost is smaller than the shortest path cost, increased by 30%. The choice of this percentage is highly dependent on the constellation. Optimisation of this percentage is subject to further investigation.
2. If the path chosen in the last time interval is not an element of the present set, or its cost is not acceptable, the best path of the set is chosen.

FD Algorithm   The FD algorithm splits the load to different paths according to the path length given by a flow dependent metric. It continuously adapts this load splitting following the changes in path length, trying to minimise the cost function, which is adaptive to the transmission delay, as given below[53]:

$$D = \frac{1}{\gamma} \sum_{(i,j)} \frac{f_{ij}}{C_{ij} - f_{ij}} + p_{ij}f_{ij} \qquad (4.7)$$

where p_{ij} is the transmission delay on link (i,j), f_{ij} is the flow on link (i,j), C_{ij} is the capacity of link (i,j). The length of the link is taken equal to the derivative:

$$\frac{\partial D}{\partial f_{ij}} = D' \qquad (4.8)$$

Table 4.10 Traffic levels

Traffic level	1	2	3	4	5	6	7	8	9	10
Packets/s	230	450	669	889	1108	1328	1548	1767	2206	4402

The FD routing algorithm that we have adopted for the selected network has been tested for terrestrial networks giving very good performance [36]. Assuming a network of N nodes, let W be the set of origin–destination pairs w. For each pair w, a number of distinct paths N_p connect the origin to the destination node. The flow of each path is denoted by x_p and the resulting vector $x = \{x_p\}$ corresponds to the network routing pattern. The objective of the routing algorithm is to find a routing pattern x that minimises the cost function D. The algorithm iterates deviating flow from non-optimal to optimal paths until the routing pattern is optimised. This deviated amount is adjusted through a parameter called step size and denoted by $a_s \in [0,1]$. For every iteration of the algorithm the value of a_s is adapted according to the following equation:

$$a = \min\left[1, \frac{\sum\limits_{(i,j)} (\bar{f}_{ij} - f_{ij})D_{ij}'}{\sum\limits_{(i,j)} (\bar{f}_{ij} - f_{ij})^2 D_{ij}''} \right] \tag{4.9}$$

The application of Eqs. (4.1–4.3) assume that the arrival pattern is a Poisson process.

Numerical Results
The performance of the three different routing algorithms have been compared through two different simulation techniques: (a) non-real-time and (b) real-time implementations. For non-real-time simulation, comparisons are concentrated on the convergence capability of different algorithms and the mean delay. For real-time simulation, analysis is made at the packet level such that every single packet uses the routing pattern produced by the above routing algorithms. The routing pattern is updated every time period equal to T_r

Non-Real-time Simulation Figure 4.21 illustrates the iterations required for the convergence of the algorithm with the mean load of the system as the parameter under symmetric load and symmetric distribution of the Earth stations. The number of iterations is found to be high for heavy load but convergence is achieved in every case.
Figure 4.22 shows the mean delay vs. the number of iterations. Two cases were studied: (a)

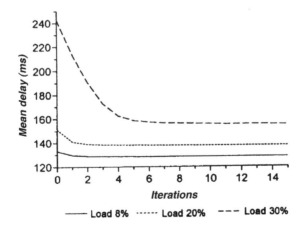

Figure 4.21   FD convergence for symmetric network load

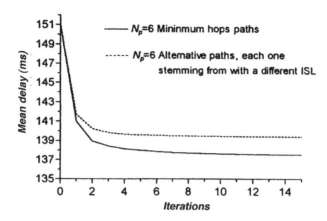

Figure 4.22 FD Convergence according to N_p for different selection criteria

any origin–destination pair uses six alternative paths, the minimum hop paths selected as in Figure 4.22 and (b) any origin–destination pair uses six alternative paths, each one stemming from a different ISL leading to six disjoint paths. The performance is better in the first case leading to the conclusion that the inter-satellite links may not be used at the time.

Figures 4.23 and 4.24 compare the performance of the three algorithms in terms of the mean delay vs. both balanced and unbalanced load. As can be seen from the result, the Dijkstra Algorithm fails at heavy load even under unbalanced load situation, which is the permanent status for satellite constellations. The adaptive Dijkstra Algorithm performs better than the ordinary Dijkstra Algorithm. However, the k-shortest path FD out-performs the others.

Figure 4.25 examines the performance of the three algorithms under different balanced traffic load. Both uniform and non-uniform distribution of Earth stations have been considered. Under heavy load condition, the simple Dijkstra Algorithm fails to perform. The other two algorithms present similar behaviour under both uniform and non-uniform Earth stations distribution. The mean delay increases slightly as the load increases.

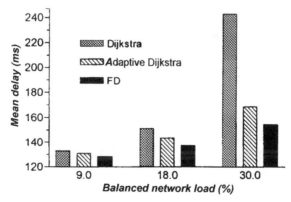

Figure 4.23 FD, Dijkstra and adaptive Dijkstra according to different amounts of balanced network load

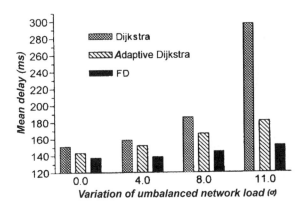

Figure 4.24 FD, Dijkstra and adaptive Dijkstra for the case of unbalanced network load according to variation, with mean network load

Real-time Simulation   For real-time simulation, both Poisson and Self-Similar input traffic are considered. Balanced network load and uniform distribution of ground stations are assumed. The network uses $N_p = 6$ paths for every pair w. The values presented in the results are measured in timeslots, which are equal to the duration for the transmission of an ATM cell.

Figure 4.26 shows the mean delay vs. time slot for the three routing algorithms. Poisson input traffic with mean value of 0.4 packets/timeslot is assumed. During the time interval of 10,000 timeslots, all three algorithms manage to route the traffic and demonstrate stable performance. The FD algorithm results in the lowest mean delay with a maximum routing interval $T_r = 100$. This implies that the FD routing algorithm out-performs the other two algorithms, even the routing update is 10 times less frequent than the others. However, if $T_r = 50$, the FD algorithm shows practically no improvement since the Poisson environment is affordable. However, the performance of the adaptive Dijkstra degrades significantly with

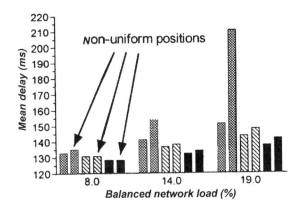

Figure 4.25 FD, Dijkstra and adaptive Dijkstra for balanced network load and non-uniform distribution of Earth stations

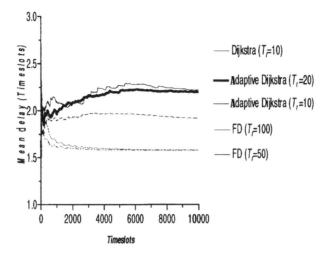

Figure 4.26 Real-time simulation for the Poisson case for the FD, Dijkstra and adaptive Dijkstra with link flow algorithms with mean input load 0.4 packets/timeslot

an increase of the routing interval ($T_r = 20$), which is similar to that of the simple Dijkstra Algorithm.

Figure 4.27 shows the performance of the three algorithms using Poisson input traffic with mean value of 0.7 packets/slot. The results show that the Dijkstra and adaptive Dijkstra Algorithms fail to route the traffic but the FD conserves a very low mean delay value by splitting the input traffic into different paths. This indicates the capability of the FD algorithm in exploiting the resources of the network.

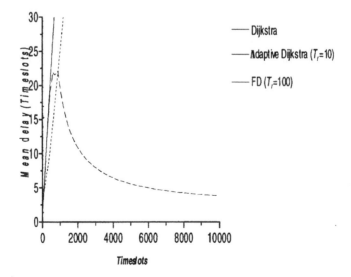

Figure 4.27 Real-time simulation for the Poisson case for the FD, Dijkstra and adaptive Dijkstra Algorithms with mean input load 0.7 packets/timeslot

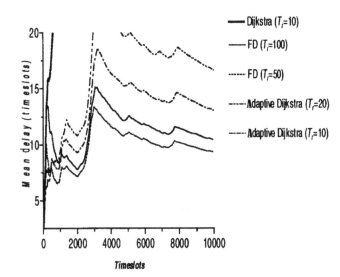

Figure 4.28   Real-time simulation for the self-similar traffic for the Dijkstra, FD and adaptive Dijkstra algorithms with a mean input load of 0.4 packets/timeslot.

However, the results using self-similar traffic differ significantly. Figure 4.28 shows that there is a significant degradation of the system performance even if the mean load traffic is equal to 0.4 packets/timeslot. The Dijkstra Algorithm presents unacceptable behaviour since the mean delay increase monotonically until packets begin to drop. The other two algorithms manage to route the traffic. The self-similar traffic is characterised by its bursty nature, which is the worst event for any routing algorithm due to the unpredictable peak value. The FD algorithm results in better mean delay value. However, the performance of the algorithm is no longer stable. Nevertheless, the routing interval for FD is still bigger than the other two algorithms.

Figure 4.29 compares again the performance of the FD and adaptive Dijkstra Algorithms using self-similar traffic of mean traffic load equal to 0.7 packets/slot. The FD algorithm still performs under heavy traffic load condition. The adaptive Dijkstra fails even when the routing interval is halved. Generally, the frequency and the policy of the routing trigger are important parameters for the design of the system, which deserves further investigation.

Case Study III
The third case study looks from a higher level standpoint at the mechanisms forming link state routing algorithms. The simulations are performed at connection level and the accent is also put on signalling aspects.

Issues in LEO Routing
Figure 4.30 presents the organisation of a network comprising a LEO system and a terrestrial system, each interconnected via a Gateway (GW). The LEO system is constituted of mobile and fixed User Terminals (UTs) and a satellite constellation with ISLs. Ground entities in the LEO system (UTs and GWs) access the satellites via Up/Down Links (UDLs). In the terrestrial system, UTs are connected to GWs via Terrestrial Network Links (TNLs).

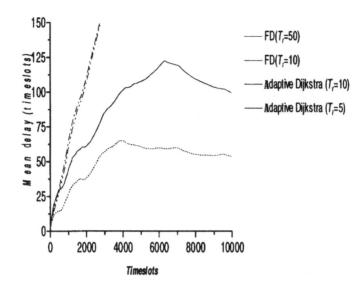

Figure 4.29 Real-time simulation for the self-similar case for the FD and adaptive Dijkstra algorithms with mean input load 0.7 packets/timeslot

Examples of TNL routing algorithms are Routing Information Protocol (RIP), Open Shortest Path First (OSPF), Intermediate System to Intermediate System routing (IS-IS), Private Network to Network Interface routing (PNNI) or Darpa Packet Radio routing [29]. Routing in the UDL segment can be categorised into up link and down link routing for selecting the first and last satellite in the route. UDL routing is trivial when one satellite is visible at a time from a UT or GW. Different policies for UDL routing exists [23]. ISL routing consists of computing a path in the constellation from the first to the last satellite chosen by UDL routing. Although UDL and ISL routing were presented as separate processes, it is possible to merge them to support end-to-end integrated routing [22].

The two next subsections present two characteristics that make LEO constellations special from a routing point of view: the dynamics of the topology and the regular structure of the network topology.

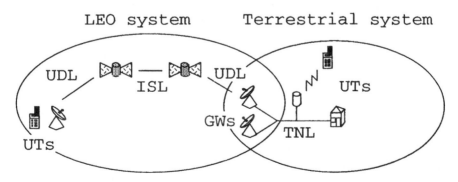

Figure 4.30   Terrestrial, UDL and inter-satellite link segments

Dynamics of the Topology

The topology of a LEO constellation is varying over time. Deterministic changes in the topology are a result of the satellite's motion. These changes occur as ISL and UDL are switched on and off periodically due to pointing, acquiring and tracking constraints. Deterministic changes can be processed in advance by the routing algorithm. Non-deterministic changes are a consequence of an unpredictable degradation of UDL quality, a failure in an equipment or a change of the traffic distribution.

Deterministic changes are almost extinct in terrestrial routing. Most of the changes are non-deterministic such as a router failing or a switch being congested. Non-deterministic changes are likely to happen often in the constellation, especially UDL fading, due to radio propagation impairments. As a result, the availability of a route in the LEO environment is a random variable.

As far as deterministic changes are concerned, decentralised algorithms are favoured. Indeed, it is possible to reflect the changes internally in each router without the need of distributing any information through the network. It is not the case for distributed algorithms where signalling is required for updating the routes. Considering the frequency of predictable changes (for Iridium the average lifetime of an UDL is 7 min without considering spotbeam handovers, while the average lifetime of a cross-plane is 33 min), these changes can cause a significant amount of signalling traffic. It is a strong point favouring decentralised algorithms.

A predictive telephony traffic model [27] can be used to predict non-deterministic changes caused by traffic variations. Unfortunately, as soon as multi-service data communication is addressed, it is difficult to establish a precise model. Firstly, the diversity of services, each with different communication requirements, make it difficult to provide a global estimate taking all services into account. Secondly, it is difficult to devise a realistic model when the traffic intensity is correlated to events occurring unpredictably in the world, such as natural disasters or political events. Predictive models should not be used without adaptive routing in order to respond to unpredictable states of the network.

Regular Structure of the Network Topology

Satellites are usually designed with symmetrical connectivity capabilities, therefore, constellations have a regular geometry. On the contrary, terrestrial networks are seldom regular, since they grow gradually through the addition of equipment or the merging of networks. Additionally, the connectivity of a LEO network which is stable, is known in advance. This is not the case of terrestrial networks where new nodes are added and removed as time goes by.

This paragraph introduces the concept of route diversity in the constellation – it is similar to satellite diversity. Given a source satellite, the route diversity with respect to a destination satellite can be defined as the number of disjoint paths (i.e. not sharing common links) between the source and the destination satellite. This measure is averaged over the orbital period. Furthermore, only paths not exceeding the shortest path length by τ percent are taken into account. The length of the path is defined as the sum of the propagation delays for all ISLs. The average route diversity of a satellite is defined as the average of the route diversities between a satellite and the other satellites in the constellation.

Disjoint paths makes the route diversity an indicator of how much traffic can be routed between two satellites for a given inter-satellite capacity.

The constellations considered in this paper are a polar constellation, namely Iridium and an inclined constellation called M-Star2 with the same number of satellites as Iridium. The

Table 4.11 Constellation characteristics

Parameters	Iridium	M-Star 2
Type	Polar	Inclined
Orbits	11	11
Sats/orbit	6	6
Altitude	780 km	1350 km
Inclination	86.4°	47°
Phasing factor	3	5
ISLs	Dynamic	Permanent
Seam	Between first and last orbit	No seam

constellation characteristics are shown in Table 4.11. M-Star2 inter-orbit ISLs are permanent while in Iridium they are switched off at the polar regions ($\pm 90° \pm 30°$). Furthermore, Iridium inter-orbit ISLs across the seam are permanently disabled.

Figure 4.31 shows the evolution of the average route diversity according to time. The satellite chosen for measuring the diversity is the first satellite of the third orbit in order to avoid a satellite close to the seam, whenever a seam exists. Paths are accepted up to $\tau = 50\%$ longer than the shortest path.

M-Star2 features a better diversity than Iridium with less variations over time. This is due to the permanent ISLs of M-Star2. The improved diversity of M-Star2 is explained by the absence of a seam. Indeed, given two satellites it is possible, from the first satellite to reach the second satellite either going westbound or eastbound. In Iridium, because of the seam, one of the paths has to go through the polar region, hence being longer. This path is possibly too long to be taken into account during the evaluation of the diversity.

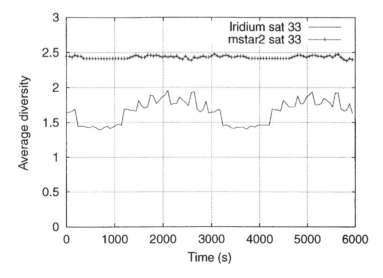

Figure 4.31 Comparison of the route diversity for Iridium and M-Star2 ($\tau = 50\%$)

The diversity is an indicator of the availability of alternate routes when the network becomes congested. The higher the diversity, requests will be blocked due to congestion. The diversity depends on the constellation connectivity. Therefore, the constellation design impacts the routing efficiency.

In the following paragraphs we compare the performance of static and adaptive routing algorithms for Iridium and M-Star2.

Static vs. Adaptive Routing Algorithms
The comparison is performed in the following example: twenty-three gateways homogeneously distributed on the ground and generating connection requests according to a Poisson law with exponentially distributed length. The connections from each gateway are randomly designated to other gateways.

The routing algorithm computes the shortest path in term of delay. A satellite introduces a fixed 10 ms switching delay while an ISL adds a delay equal to the propagation delay. Each satellite has a switching capacity of 100 connections; ISLs have a capacity of 25 connections; UPL have a capacity of 100 connections. Saturated satellites and ISLs are not taken into account during routing.

Adaptive routing is implemented by having the satellites distributing state information (in terms of unused capacity) every 30 s. In static routing, the distribution happens only once at the beginning of the simulation when the traffic load is zero (unused capacity is equal to maximum capacity).

In Figure 4.32 the connection blocking probability is compared for static and adaptive routing under varying loads. The load is expressed in Erlangs, originated from one gateway.

Adaptive routing improves significantly the connection blocking probability by using alternate routes when congestion occurs. Static routing in M-Star2 performs better than adaptive routing in Iridium. Given the blocking probability a larger traffic can be supported with adaptive routing. M-Star2 outperforms Iridium as a result of its higher route diversity.

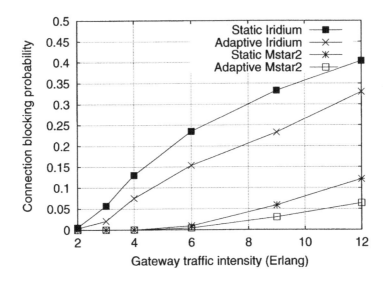

Figure 4.32   Connection blocking probability for static and adaptive routing

Table 4.12 Comparison of the two causes for blocking according to their percentage of occurrence with adaptive routing (Iridium case)

Traffic intensity	Blocking probability	Ratio of routing failures (%)	Ratio of set-up failures (%)
3	0.02	0.00	100.00
4	0.07	2.00	98.00
6	0.15	6.00	64.00

Adaptive routing has another interesting benefit. A connection might get blocked for two reasons [28]. Firstly when the routing algorithm is not able to find a route – this is a *routing failure*. Secondly, should no routing failure occur, the connection is set-up by having a call signalling message going down the route and making resources reservation in each node. If no resources are available on the route, the connection cannot be set-up and this is a *set-up failure*. Set-up failures are caused by obsolete network state information used for the routing algorithm. Routing and set-up failures are not differentiated from a user point of view. However, from an operator standpoint, routing failures are less expensive than set-up failures. Indeed, set-up failures result in wasted network resources with unsuccessful call signalling. Table 4.12 shows the evolution of the ratio of routing and set-up failures for blocked connections at different gateway traffic intensities for Iridium. The ratio of set-up failures decreases as the load increases. With static routing, blocked connections are always due to set-up failures.

The dimensioning of the UDL is a sensitive parameter. In the above example, the UDL capacity is equal to the sum of the satellite ISLs capacity. Nevertheless, if the UDL capacity is smaller, most of the congestion takes place in the UDL segment and the blocking probability improvement of adaptive routing becomes marginal.

Isolated vs. Non-Isolated Routing
The static algorithm presented in the previous section can be transformed into an isolated adaptive algorithm by including in the algorithm the traffic intensity going through the satellite performing the routing computation. Simulations, not reported here, showed that there is no significant difference between the adaptive isolated version of the algorithm and the static version. The focus is therefore put on non-isolated algorithms and the distribution of network state information.

As seen from the results of section static vs. adaptive routing algorithm, distributing information tends to lower the connection blocking probability. Nevertheless distributing information to all nodes generates signalling traffic. It is possible to keep the signalling traffic low by restricting the number of distributions. Unfortunately, it affects the routing algorithm since routing will be performed with obsolete information. A balance must be achieved between routing accuracy and signalling traffic.

Satellite constellations have an advantage over common terrestrial networks since the topology is mostly regular and known in advance. It is therefore possible to design distribution schemes having a low complexity in terms of number of packets sent in the network.

Two distribution schemes are evaluated. In the *Periodic scheme*, each satellite distributes information periodically, as specified by the *Broadcast Period* parameter. In the second

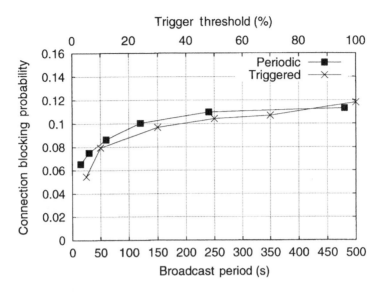

Figure 4.33 Connection blocking probability for Periodic and Triggered distribution schemes at different periods and trigger thresholds (Gateway traffic intensity = 4 Erlangs)

scheme, called *Triggered*, a satellite distributes information each time its available capacity (or the available capacity of one of its ISL) varies from an amount exceeding a *Trigger Threshold*. The threshold is a percentage of the initial available capacity. One might expect that the Triggered scheme will help decreasing the signalling traffic since information is only distributed upon significant change of the network state. The two schemes will be compared for a gateway traffic intensity of 4 Erlangs. Different values are assigned to the Broadcast Period and Trigger Threshold. Finally, both the connection blocking probability and the number of distributions are used to quantify the trade-off between route accuracy and signalling load.

Figure 4.33 shows the evolution of the connection blocking probability with respect to the Broadcast Period (bottom *x*-axis) and the Trigger Threshold (top *x*-axis). The periodic and triggered schemes display similar trends.

Table 4.13 compares the number of state information broadcasts per sample period of 1000 s using the periodic and triggered schemes. The comparison is presented for three values of the connection blocking probability. The triggered scheme prevents consuming capacity in useless signalling of state information.

Table 4.13 Number of state information broadcasts per sample period of 1000 s

Blocking probability	Periodic	Triggered
0.07	3300	2734
0.08	1650	1387
0.1	550	128

Although not reported here, simulations with the M-Star2 constellation provided similar results.

Pre-Computed and On-Demand Routes

Until now, the routes were computed on-demand. A routing algorithm with pre-computed routes is now evaluated. All satellites compute periodically (*pre-computation period*) the routes to all destinations. The algorithm is adaptive (with network state information distributed every 30 s) and uses the end-to-end delay as optimisation metric. The connection blocking probability of the pre-computed scheme is evaluated as the pre-computation period varies. The traffic intensity from each gateway is equal to 4 Erlangs.

Figure 4.34 shows that the connection blocking probability quickly degrades as the pre-computation period increases. It is a result of the discrepancy existing between the current network state (at connection set-up) and the network state at the time the routes are computed. Such a discrepancy is caused by the evolution of the constellation geometry and also changes to the traffic load distribution. Figure 4.33 also presents the evolution of the connection blocking probability for Iridium with motionless satellites, hence removing the first cause of discrepancy. It turns out that variation in the constellation geometry is the main reason for the degradation of the pre-computed scheme performance. Although route pre-computation is an efficient strategy for terrestrial networks, it is no longer such with satellite constellations.

An additional issue is raised due to the number of routes that has to be computed in advance. Given a pre-computation period equal to 240 s, the pre-computed scheme requires 17 route computations per second, while it drops to 0.5 for the on-demand scheme. However, as far as the on-demand scheme is concerned, the route computation rate depends on the network load.

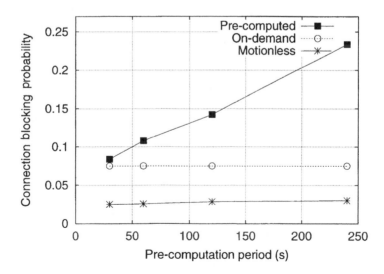

Figure 4.34 Connection blocking probability for the pre-computed scheme at different pre-computation periods (GW traffic intensity = 4 Erlangs). The horizontal line with circle shaped points corresponds to on-demand blocking probability

Centralised, Decentralised and Distributed Routing

This subsection covers only decentralised and distributed routing. Since the use of centralised routing is prone to create bottlenecks and features a single point of failure, it is discarded. As mentioned earlier, adaptability to predictable changes is more convenient for implementing decentralised routing algorithms. The same conclusion was drawn for on-demand routes. The issues raised by routing loops are also a strong point against distributed algorithms. On the other hand, decentralised routing algorithms require nodes equipped with significant computing power. Whether the algorithm is meant to be part of an on-board router must be taken into account. Considering the evolution of today's VLSI, both from the gate density and power consumption point of view, decentralised algorithms are preferred although the distributed approach is not completely discarded.

Connection-Less and Connection Oriented Modes

As mentioned before, QoS routing is barely possible in connection-less environments. Unfortunately, supporting connection oriented mode in LEO networks creates a problem. Since the links are switched on/off periodically, connections going through these links must be re-routed. This event is called a connection handover. In connection-less environments, no connection is required, events such as links switching on and off are directly reflected in the routing tables. Re-routing the connection is costly and is the source of jitter. Such problems are not wanted, especially for real-time services. As a result, a routing algorithm must aim at minimising the impact of handovers. Soft handover, incremental re-routing and handover avoidance are techniques that should be included in the routing algorithms.

Existing Routing Algorithms for LEO Constellations

Routing for LEO networks is covered in [21,24,25,26,27]. All but [25,26] use static or pseudo-adaptive, pre-computed routing. In some cases the routing algorithm is adaptive to connectivity changes but not to the traffic [24]. The constellation topology evolution is in discrete steps, during which it is considered quasi static such as in [21]. Although convenient, this approach has a drawback. Since all route updates occur according to the discrete step, some updates might be realised too early and others too late. Reference [25] considers on-board distributed algorithms using pre-computed routes (every 60 s) and another considers decentralised routing located in the gateways [26].

Signalling is covered only in [25] and is identified as a major issue. Signalling is not to be neglected because the dynamic topology calls for a careful design of the signalling protocol especially with decentralised algorithms [26].

Handovers are taken into account in [24,26,27], the other authors assume a connection-less environment.

None of the algorithms supports QoS routing. Reference [25] have to be significantly modified for QoS routing. This is an unfortunate characteristic of distance vector algorithms which are not well suited to on-demand routing. Referring to the remarks made previously, we are sceptical about the suitability of distributed algorithms for QoS routing.

To summarise, two contributions [26,27] are the closest matching algorithms, in respect to the identified needs. These algorithms can be traffic-adaptive if provided with the appropriate signalling. They also support the connection oriented mode and take into account handovers. Furthermore, they are decentralised, easing the implementation of on-demand route computation. This aspect is crucial for future support of QoS routing. Unfortunately the authors did not address signalling issues related to the gathering of network state information.

4.3 Congestion Control

4.3.1 Overview

Congestion control is an important issue in ATM networks, which offer the possibility of multiplexing different types of traffic on a single link. Each traffic type may require different QoS (cell delay, cell delay variation, cell loss ratio...). To meet those different and independent requirements, traffic management should be applied, especially congestion control. As an example, the application of a suitable congestion control mechanism helps avoid excessive delays for delay-sensitive traffic, where excessively delayed packets become useless.

This is an even more important and fundamental concept in the satellite environment, where propagation delays are much higher than in fibre-optic terrestrial networks. There are two types of congestion control mechanisms: the predictive mechanisms; and the reactive mechanisms.

Predictive congestion control aims to prevent congestion by taking appropriate action. It acts before the network is overloaded. Call Admission Control (CAC) is an example of preventive control. In CAC, traffic parameters are negotiated before the connection is established. The usable bandwidth of the connection is based on this negotiation, called traffic contract.

Reactive congestion control, contrary to the predictive control, reacts to an already existing congestion state. It attempts to recover from this congestion state. The efficiency of such control depends widely on the efficiency of the applied feedback mechanism.

Rate-based control is a reactive congestion control. It is an end-to-end control using a feedback loop to control the cell emission rate of each connection. This control scheme was chosen by the ATM Forum as the Available Bit Rate category (ABR) congestion control.

4.3.2 The LEO Satellites Network Environment

As already mentioned, congestion control is more important in the LEO satellite environment than in the terrestrial networks. This is due essentially to the satellites' limited bandwidth. However, the congestion control mechanisms applied in terrestrial networks, are no longer sufficient and well-adapted to the LEO Satellite environment.

In its movement around the Earth, a LEO satellite will experience different levels of traffic load. This is due to the differences in term of geographical, demographical and 'technological' characteristics of the covered zones. It is obvious that the number of sources connected to the satellite, whilst it is located above the ocean, is more reduced than whilst it is located above a continent. The same can be said if the satellite is above New York, where the number of potential active users is high, or if it is above London. The difference becomes more important if it is above some African country where there are few or non-existent users. Therefore, due to its continuous movement around the Earth, the satellite could experience very significant traffic fluctuations.

This traffic non-uniformity is a fundamental point making the congestion control mechanisms used in the terrestrial networks no longer adaptable to the LEO satellite networks. The

non-uniformity is not the only element. The high propagation delay in this satellite environment, compared to the terrestrial one, affects the accuracy of the information carried in the
feedback loops.

4.3.3 Network Scenario

In our work, we consider a LEO constellation with OBP and using ISLs. It was assumed that
at each time, at least two satellites are simultaneously visible to a point in the Earth. The ABR
category of service of the ATM Forum was focussed on and a reactive rate-base congestion
control mechanism, namely the explicit-rate mechanism, was chosen.

In the explicit-rate mechanism, Resource Management (RM) cells explicitly inform the
source of its desired new rate. The destination end-system calculates this explicit rate depending on the path congestion. However, congested switches can reduced this rate, or generate,
under extreme congestion, an RM cell to send immediately to the source. These RM cells are
called out-of-rate RM cells.

The relatively high propagation delays characterising the satellite communications,
increase the probability that the source received an inaccurate rate notification. Due to the
satellite motion, the state of the network is changing continuously. Virtual Source/Virtual
Destination (VS/VD) scheme was proposed to solve delay problems in WAN terrestrial
networks. It consists of the segmentation of the control loop. The switches become both
virtual sources and virtual destinations. As a destination end-system, a switch turns around
the RM cells to the sources and as a source it generates an RM cell for the next segment. This
approach reduces the delay required to adjust the source rate. On the other side, it increases
the number of RM cells in the network. This is a real drawback in the satellite environment
where the bandwidth is scarce.

A simplification of the original explicit-rate control consists of using exclusively out-of-
rates RM cells. The source receives an RM cell only and directly from the congested switch.
This simplification is reducing the power of the original scheme, but could be adapted to the
characteristics of the satellite environment. However, two problems arise. The first is the
priority of the out-of-rate RM cells. The Cell Loss Priority (CLP) of those cells is set to 1,
which means low priority. If only out-of-rate RM cells are exchanged, this priority should be
set to the higher one, namely 0. The other problem is the number of allowed out-of-rate cells,
which is limited to 10 cells per second per VC.

In our work, it was suggested to compare the performance of a LEO satellite network with
the original explicit-rate, the VS/VD and the simplified explicit-rate schemes. This last
scheme could be enhanced or abandoned depending on the results.

Traffic non-uniformity is a more general problem. It creates important fluctuations in the
satellite load. Furthermore, it may create substantial differences in traffic load between
satellite neighbours. A satellite experiencing congestion could have neighbours not using
all their capacities. To assure a fair load distribution, satellites should periodically exchange
information about their load state. Their service areas could be adjusted depending on that
information. A satellite may extend its service area to the benefit of a heavily loaded
satellite neighbour. The last one will see its service area collapsing in consequence and
its traffic load reduced. This approach was inspired from works done in CDMA LEO
satellite networks.

This proposed approach is a predictive one. A reactive one is to balance the traffic load only when congestion happens. This reduces the flow of load state information exchanged by the satellites. Future work will be carried out to investigate the performance of the proposed technique.

4.4 Multicast

4.4.1 The Broadband Constellations

Brief descriptions of the main contenders in the LEO broadband constellation arena are discussed in this section.

4.4.1.1 Teledesic

Announced in 1994, Teledesic is the first and most ambitious of the LEO broadband constellations. Originally planned as 21 near-polar orbital planes of 40 active satellites and 4 in-orbit spares per plane at an altitude of 700 km [37], in early 1997 it was scaled back to 12 near-polar planes of 24 active satellites and 3 in-orbit spares per plane, at an altitude of 1350 km (Figure 4.35).

Each Teledesic satellite in the original constellation was proposed to have eight ISLs in the 60 GHz band, for its two nearest neighbours in each of the four cardinal directions, forming a complex cylindrical mesh network, a topology described as 'geodesic'. The redesign now uses optical ISLs, rather than phased-array antennas, for increased capacity. Constellation topology is discussed in more detail in [39].

Teledesic ground terminals will provide data rates from 16 kbps to 2 Mbps in 16 kbps increments, while gateway 'GigaLink' terminals between Teledesic and ground networks have data rates from 155 Mbps (OC-3) to 1.2 Gbps (OC-24). Each satellite acts as a fast-packet switch, using a proprietary fixed-length 512-bit packet format with a proprietary connectionless adaptive-routing protocol.

Ka-band frequencies were allocated to Teledesic at the 1995 World Radio Conference (WRC) and the US Federal Communications Commission (FCC) granted a licence in March 1997. Construction has not yet begun, but service is scheduled to begin in 2002.

4.4.1.2 SkyBridge

In February 1997, Alcatel announced its SkyBridge constellation, a redesign of its earlier Sativod proposal and filed an application for frequency with the FCC (Figure 4.36) [40].

SkyBridge is intended to share Ku-band with existing GEO satellites. It uses two overlapping offset constellations of 32 satellites each (each being four planes of eight active satellites per plane, all at an altitude of 1457 km) to give a choice of satellites to ground stations. For each satellite pair, the satellite furthest from the centre of a GEO footprint that satellite footprints overlap, is used for minimum interference with the GEO satellite.

SkyBridge does not use intersatellite links. Its satellites are planned to act as in-orbit 'bent-pipe' transponders, or amplifying repeaters, forwarding traffic to local gateways. This can therefore be thought of as a ground network with a space-based component, rather than as a space-based network, as all the switching fabric is ground-based.

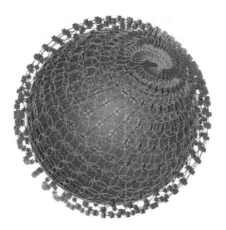

a. Original design

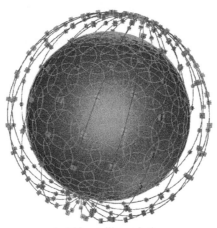

b. 288-satellite redesign

Figure 4.35   Teledesic satellites and ground coverage rendered by SaVi [57]

User terminals offer data rates from 16 kbps to over 60 Mbps, in increments of 16 kbps, using an ATM cell-switching interface.

Alcatel plans to deploy one 32-satellite constellation by 2001 and the second by 2002.

4.4.1.3 Celestri

In July 1997 Motorola announced Celestri and filed an application for a Ka-band licence with the FCC (Figure 4.37) [41].

Celestri combines a LEO constellation (similar to the M-Star constellation also proposed earlier by Motorola) with a GEO constellation (developed under the name Millennium). The LEO constellation has nine planes, each containing seven active satellites and one spare, at an altitude of 1400 km.

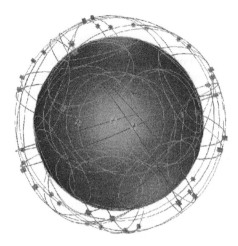

Figure 4.36   SkyBridge satellites and ground coverage rendered by SaVi [57]

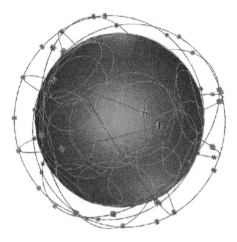

Figure 4.37   Celestri LEO satellites and ground coverage rendered by SaVi [57]

Each satellite has six optical intersatellite links of 4.5 Gbps capacity, forming a toroidal mesh network around the Earth and an on-board ATM switch with throughput of 17.5 Gbps. There are no ISLs connecting LEO and GEO satellites.

Communication rates to terminals are either from 64 kbps to 2 Mbps in increments of 64 kbps, or at the fixed rates of 10 Mbps, 16 Mbps, 51 Mbps (OC-1) or 155 Mbps (OC-3). Protocol adapters for transparently carrying various network protocols over the Celestri network are to be included in terminals, presumably to tunnel these protocols over the underlying ATM infrastructure.

Motorola has since become involved in the Teledesic proposal and Celestri as described is unlikely to be launched.

4.4.2 Multicast Technology

One-to-one end-to-end connectivity across an internetwork can be accomplished by circuit-switching or by a virtual circuit. For group applications, where more than two users simultaneously exchange messages and maintain state, the number of circuits required would increase rapidly as the number of users increases. To prevent applications from needing to know about all users in the group or needing to be responsible for maintaining all these circuits and to decrease network load, we require multicast, which is the efficient emulation of broadcast within the constraints of a network environment.

Multicast allows a source to simultaneously send data to all users on the internetwork interested in receiving the data, but in a more efficient manner than simply flooding the entire internetwork with unnecessary broadcast packets.

The set of all interested hosts form a multicast group and group management is an internetwork function. Users not in the group do not see the data (perhaps because there is no need for the data to be sent across their network) or, if they do see it, they discard it.

To communicate the data efficiently to all users in the group, the internetwork must set-up a spanning tree, along which the messages can be sent and replicated at tree branches.

4.4.2.1 Multicast in Existing Network Stacks

The IP implements multicast groups as part of the network address space, using reserved Class D addresses, where there is a one-to-one relation between a multicast group and a D address at any given time. Considerable work has been carried out on the implementation of a variety of multicast protocols across the Internet. IP multicasting protocols rely on the Internet Group Management Protocol (IGMP) to manage multicast groups [42]. There is a wide choice of available multicast routing protocols, such as the Distance Vector Multicast Routing Protocol (DVMRP) [43] and Multicast Open Shortest Path First (MOSPF) [44].

These protocols can be split into two main groups, according to the assumptions that they make about group distribution and network traffic.

Dense-mode multicast protocols assume that multicast group members are densely distributed throughout the internetwork, i.e. that many subnetworks contain at least one group member and that internetwork capacity is plentiful. These are data-driven, in that construction of the multicast tree begins top–down from the source outward and data on the state of the tree is propagated to all routers. The multicast trees constructed allow data to travel in one direction, from source to group, emulating broadcast.

For full group-to-group communication, a multicast tree must be set-up from every user; this scales badly for large groups and imposes joining and leaving overhead and delays. DVMRP and MOSPF are dense-mode.

Sparse-mode multicast protocols assume that group members are sparsely distributed and that capacity is constrained. In sparse mode, the multicast tree is receiver-initiated, i.e. a router becomes involved in the construction of a multicast distribution tree only when one of the hosts on its subnet requests membership. Core-Based Trees (CBT), with a single central router from which the tree branches out in all directions, have been suggested for groups where there are many active senders within the group, allowing multi-way communication over a single tree. This effectively minimises the amount of multicast state routing information that needs to be stored and the number of routers involved. This single-tree approach differs from the (source, group) pairings of DVRMP and MOSPF and scales better for large networks, with minimal joining and leaving overheads. Protocol-Independent Multicast-Sparse Mode (PIM-SM) constructs a multicast tree around a chosen router, called a rendez-vous point, similar to the core in CBT [45].

Given the constraints of a satellite constellation network with intersatellite links, a sparse-mode multicast protocol would appear to make best use of limited, expensive, capacity. However, as the satellites move in their orbits it would become necessary to periodically move the core of the tree and this would also require support for native multicast routing in the satellite on-board switches.

With its circuit-switching focus, ATM did not originally have the concept of multicast and the closest it came to multicast was the concept of a multicast server, where virtual circuits from all receivers would connect to a single source. This scales extremely badly in large networks with many users, as the links adjacent to the server soon run out of capacity and available Virtual Path Identifiers (VPIs) to allocate to connections.

Work is now taking place to add multicast support to ATM, e.g. SEAM, which uses a single core-based tree and combines ATM VC to minimise VC switching overhead [46]. However, IP multicast is a considerably more mature area. It is likely that much of the traffic over ATM networks, including satellite ATM networks, will be generated by applications using IP and tunnelled through ATM.

4.4.2.2 Internetworking with Multicast

When networks on an internetwork are not multicast-aware, multicast messages can be *tunnelled* through them by *encapsulating* (and if necessary, fragmenting) the multicast messages within network messages native to that network.

This is the strategy adopted by the MBone, or Multicast Backbone, where isolated IP-multicast-aware networks are internetworked together across multicast-unaware networks by tunnelling IP multicast messages through ATM circuits between the multicast islands. This forms a virtual network layered on top of the physical network.

However, tunnelling gives a false picture of the latency of a connection as measured by a Time-To-Live (TTL) or hop counter, since the header containing the count value is encapsulated and is not decremented in the tunnel, no matter how long the tunnel is. All tunnels appear the same length to the packets being tunnelled through them.

Tunnelling does not make optimal use of the intervening network. The tunnel must often be set-up manually and the tunnel endpoints cannot be moved easily. The MBone lacks flexibility.

Adapting the TCP/IP and ATM network stacks to each other in order to increase through-put when IP is layered over ATM, has been an area of considerable interest for some time. Adaptations such as TCP Boston, where the implementations of both protocols are modified slightly in order to increase tunnelling performance, are starting to appear. [47]

As an extension of this, protocol federation of ATM multicast protocols with IP multicast protocols is one possible, albeit highly unlikely, solution to allow true IP/ATM internetwork multicasting.

4.4.2.3 Multicast Over Satellite

Given the maturity and widespread adoption of IP multicast protocols, we can expect a demand for IP-multicast-aware group applications to be used in wireless environments, including satellites.

Satellite-based videoconferencing could be accomplished by tunnelling the IP-multicast messages through the satellite gateways. However, this would require the setting up of multiple tunnelled virtual circuits between geographically-separate users, making group management difficult and using more satellite capacity than would be necessary if the satellites' on-board switches were to support IP multicast directly. The proposed commercial schemes do not carry this out.

SkyBridge treats each of its satellites as a simple transparent repeater, resulting in each satellite connection acting as a simple short one-hop ATM tunnel that is transparent to IP multicast and the problem of internetwork multicast is moved away from the satellites into the ground networks that are utilising SkyBridge satellites for connectivity.

Celestri is effectively a standalone ATM network utilising ATM switching in the ground–air interface and between satellites. Tunnelling of IP multicast packets over a changing-geometry ATM switching fabric will not permit multicasting within the constellation. The inclusion of a complementary GEO component in Celestri for broadcast applications appears to tacitly acknowledge and attempt to side-step the networking limitations of the LEO component. However, the increase in propagation delay for the GEO component is offset by the decrease in the amount of switching required.

Teledesic utilises a purpose-designed 512-bit fixed-length packet and is likely to encapsulate or tunnel even ATM across its network. This protocol tunnelling across the Teledesic geodesic mesh is likely to make implementation of IP multicast within the constellation network impossible.

4.5 Reliability

4.5.1 Background

Several new systems have been proposed in recent years for the provision of global telecommunication services based on different non-geostationary satellite constellations. The operation of the first LEO constellations designed for the provision of mobile services has demonstrated the technical viability of this challenging new approach. The modern satellites, designed to be utilized in the constellation systems, have become more and more complex in order to cope with the new demanding system requirements defined for the provision of innovative telecommunication services. Today the satellites must be considered as nodes of an actual global telecommunication network demanding for the adoption of a suitable network Operation & Maintenance policy, as applied in any terrestrial network.

The final aim of the policy is to guarantee a given level of availability for the space segment. Based on the objectives for network performance, the achievement of the target level of QoS in a given system, constitutes one of the key elements for the success (both technical and commercial) of the specific system. The introduction of proper maintainability criteria can dramatically contribute to improving the performances of the network, as perceived by the end user. Most of the methodologies for the operation and maintenance of the currently deployed ground networks, could be applied, profitably, to the ground segment of these new integrated satellite and ground networks. However, the problem of controlling and maintaining a constellation of satellites by considering the effect of the possible occurrence of partial or catastrophic failures is still quite unexplored.

The adoption of payloads with on-board regeneration, switching capabilities and on-board processing has surely increased the flexibility of the satellite use and made possible the advent of the satellite constellation systems. On the other hand, such flexibility needs to be carefully managed in order to maximise the level of performance achievable with these sophisticated and expensive switching nodes. Nowadays, the classical concept of the satellite statically interconnecting the entities involved in every specific connection, has radically evolved with the adoption of non-geostationary orbits and the related dynamics introduced by the orbital motion of the satellites. In most of these new systems, the antenna footprints on the ground move and overlap by following predefined paths directly related to the specific mechanical and radio characteristics of the satellite constellation.

The nature of the satellite makes impossible the adoption of maintenance strategies (i.e. replacement of failed parts) typical of the terrestrial environment. An additional problem can be caused by those minor failures which could affect a satellite without impacting on its full capability of properly operating. Based on the O&M criteria applied in any specific system, the satellite could not be de-orbited and should continue to operate with degraded performance. Since the design lifetime of the modern satellites is increased with respect to what happened in the past and any constellation can consists of up to 200–300 satellites the probability of having in a constellation several satellites affected by different partial failures could be not negligible (it mainly depends on the maintenance policy adopted in the constellation and normally results from some trade-off among QOS and cost/benefits). In this network scenario, the failures present in the satellites of the constellation are no more permanently associated to a specific geographical region of coverage on ground, but due to the orbital motion of the satellites, affect all the different areas serviced during specific time frames. In other words, the failure migrates across different geographical locations according to the orbital path followed by the failed satellite.

During the time frame required to replace the failed satellite (depending on business decisions it could vary from a couple of weeks up to some months), the QOS provided by the satellite network is affected and a careful design and management of a satellite constellation network should thoroughly consider these parameters in order to be able to effectively guarantee the expected level of service provided to the final end user.

For this reason it is important to identify and develop suitable methodologies able to model and characterise the behaviour of a satellite constellation in terms of Reliability. The classical definition of reliability as follows:

> Reliability is the probability that a device will operate successfully for a specified period of time and under specified conditions when used in the manner and for the purpose intended.

This definition emphasises the need to define a successful mission which implies that failures that keep the device from performing its intended mission will not occur. Reliability values are usually expressed as probabilities and failures are characterized by means of a *failure probability density function (p.d.f.), f(t)*.

The statistical reliability represents the probability that an item will survive for a stated interval, i.e. that there is no failure in the interval (*0* to *t*). This is given by the *reliability function R(t)*, defined as follows:

$$R(t) = 1 - Q(t) = 1 - F(t) = 1 = \int_{-\infty}^{t} f(\tau)d\tau \tag{4.10}$$

In the frame of the COST Action 253, a first effort has been devoted to the assessment of one possible methodology for modelling the occurrence of partial and/or catastrophic failures in the satellite of a constellation. In the following of this chapter, the main characteristics of this methodology are highlighted.

4.5.2 The Modelling and Simulation Method

The modelling methodology proposed [48,49] consists of the steps described in the following (Figure 4.38 highlights the related input and output parameters):

- Reliability modelling and off-line analysis

    1. *Development of the reliability model of each independent satellite subsystem.* The model of the satellite subsystems should be detailed at equipment level. This means that the typical failure rate does not refer to the elementary mechanic or electronic components (i.e. transistors, resistors, gates, IC's, chips, etc.) which constitute the specific equipment, but represent an overall value, derived either from a lower level modelling task or from a reliability apportioning originated by the subsystem design requirements. For models of this type the appropriate formalism is the *Reliability Block Diagram* (RBD) [50–52].
    2. *Collection of typical elementary or aggregated failure rates.* This is the logical next step to be carried out when a real system is modelled. Since activities in COST253 have not been tailored to the analysis of one specific constellation system, general assumptions on the failure rates have been adopted without affecting the general validity of the methodology.
    3. *Calculation of the reliability function for each satellite subsystem.* The calculation of the function reliability $R(t)$ can be a complicated task, depending of the complexity of the RBD model. A general purpose reliability calculation tool named Reliability Block Diagram Analyzer (RBDAN™), developed by CSELT in the frame of the participation to the Iridium project [53], has been re-used in the COST253 activity. RBDAN™ allows the evaluation of the end-to-end reliability (or availability). The obtained reliability function is the probability that the satellite subsystem is properly working, calculated for different values of the time t.

- Constellation cycle of life modelling and off-line simulation

    4. *Hypothesis about the constellation deployment plan.* The reliability functions of the satellite subsystems is equal for all the satellites of the constellation (the assumption here is that the constellation consists of replicas of the same satellite type). Because of the time required for the deployment of the constellation, the age difference between satellites is not negligible and should consequently be taken into account. This requires to make assumptions on the possible constellation deployment timing in order to accurately execute the simulation.
    5. *Hypothesis about the constellation O&M policy.* Another element affecting the age of

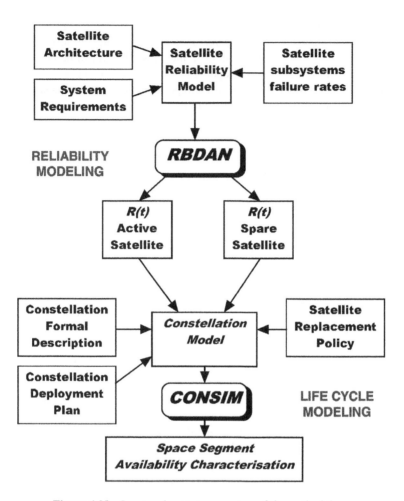

Figure 4.38 Input and output parameters of the methodology

the satellite in a constellation is related to the O&M policy for the management of the network, which identifies and replaces failed satellites [54].

6. *Simulation of one constellation cycle of life.* Once the criteria mentioned in steps 4 and 5 have been defined, a simulation can be executed in order to simulate the stochastic behaviour of the satellites which, during the constellation cycle of life, experienced partial and/or catastrophic failures. To the purpose, the preliminary version of a simulation environment, named CONSIM™ (CONstellation SIMulator), developed by CSELT [55–57] in the frame of a self-funded project and briefly described in Section 5.4 has been used.

7. *Iteration of step 6 until the target precision of result has been achieved.* In order to obtain statistical meaningful results, the simulation described in the previous step must be iterated several times.

8. *Statistical analysis of the obtained results.* The result of the several iterations can then be utilized for statistical analysis and to provide aggregate result.

9. *Identification of the constellation failure modes most likely to occur and worth analys-ing from a routing strategy standpoint.* Finally, the identification of the composite failure modes which are likely to occur in the constellation can be performed. These failure modes (in terms of failure topological configuration and probability of occur-rence) should be considered as one of the input parameters for the design and perfor-mance analysis of responsive routing algorithms to be used in the satellite constellation.

4.5.3 Payload Reference Functional Architecture and Model

In order to develop the methodology without referring to any specific system, a general reliability model of the satellite payload has been developed by adopting realistic assump-tions about its architecture. In particular:

- The communication payload is regenerative;
- It includes on-board processing and switching functions;
- Inter-satellite links are present among satellites of the same constellation;
- User access is provided by means of multi-spot coverage;
- Independent overlapping coverage using non-overlapping frequency bands and different spot contours could be present to accommodate ground terminal of different classes (i.e. small terminal and gateways);
- Coverage could be provided either by moving spots on ground (fixed pattern generated by the on-board antenna) or by fixed spots on ground (adoption of the beam steering techni-que)

The payload functional architecture as shown in Figure 4.4 is used as the reference payload architecture for the following discussion.

The architecture of any actual satellite (number of beams, RX/TX chains, presence of different types of coverage, internal architecture of the various subsystems, ... etc.) varies from system to system depending on system design choices and requirements.

By assuming that, inside the base-band OBP, each link is processed by a dedicated link processor and that the required processing capability is distributed among the base-band link processors and the base-band switch, a realistic RBD model of the satellite payload is as shown in Figure 4.39.

The RBD formalism is used to represent all the typical redundancy topologies (active, stand-by, n out of m, ...etc.) which are introduced by designers to increase the overall system reliability. The advantage of using the RBD in real engineering applications is that the modelling activity can be performed in parallel with the design progress because any RBD is already consistent (even if not really accurate) even if it represents the considered system at a very high functional level (usually at the beginning of a project). As soon as the definition of the system improves, it can be reflected in the RBD model by detailing the relevant block (i.e. defining an RBD embedded in the block) with a classical top–down approach able to provide useful feedback to the designers from the early phases of a project.

In order to properly model the cycle of life of a constellation, two different reliability models of the satellite are necessary in order to take into account the fact that, if a satellite has been positioned on the spare orbit, it has just a few of its subsystems turned on, while the remaining have not yet begun their actual operational life. The reliability model of the spare satellite consists of a subset of the complete model of the operational one and the relevant RBD is shown in Figure 4.40.

Figure 4.41 shows the reliability functions for the operational and the spare satellite.

4.5.4 Simulation of the Cycle-of-Life of a Constellation

Once the satellite reliability functions have been evaluated, the methodology requires the definition of a model for the constellation in terms of behaviour and architecture. A number of considerations applies, in particular:

- The reliability function applies for all the satellites of the constellation (i.e. the constellation consists of replicas of the same satellite type).
- Since the time required for the initial deployment of the constellation is not negligible, the age difference among the satellites must be taken into account.
- The criteria to identify and to decide about the replacement of a failed satellites must be defined.

The 'proof-of-concept' model of the constellation used has been based on the following assumptions:

Number of operational orbits	7
Total number of satellites	63
Number of operational satellite per orbit	9
Spare orbits per operational orbit	1
Spare satellites per spare orbit	1
Number of satellite types	1

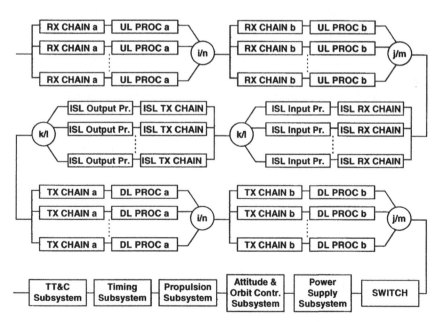

Figure 4.39   RBD model of the operational satellite

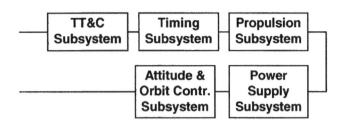

Figure 4.40   RBD model of the spare satellite

Operational satellite reliability function	as shown in Figure 4.41
Spare satellite reliability function	as shown in Figure 4.41
Initial constellation deployment time	1 year
Delay for replacement of one operational satellite	6 weeks
Delay for replacement of one spare satellite	15 weeks
Delay for spare to active transition	1 week
Probability of successful satellite in-orbit delivery	0.9
Probability of successful spare satellite switch-on	0.98
Probability of successful spare to active transition	0.95

When an operational satellite fails and no spare satellite is available in the associated orbit after a pre-defined delay, a new satellite is launched from the Earth (after de-orbiting of the failed satellite).

When a spare satellite fails after a pre-defined delay, which is longer than that in the previous case, a new satellite is inserted.

The related simulation of the constellation cycle of life has been executed by using CONSIM™. As an example of the obtained results, Figure 4.42 shows the probability of a satellite failure (not yet replaced) in a nominal orbit slot as a function of the constellation lifetime. Figure 4.43 represents the probability of having a 'service hole' in a nominal orbit slot of the constellation (i.e. failed satellite or no satellite at all).

When the methodology is applied to a real system, the obtained characterisation can be used to perform:

- Evaluation about the effectiveness of the adopted replacement policy;
- Characterisation of the availability for the space segment of a constellation;
- Global system performance evaluations;
- Identification of the most critical failure modes for the space segment;
- Analysis of the impact on the service provided of the adoption of different routing strategies able to respond to the failure events in the space network.

The last issue of the list could be very important to properly assess the system performances from the networking standpoint. In this field, an aspect affecting the QoS provided is, in fact, the routing policy when the considered system uses satellites with OBS capabilities and ISLs. Therefore, when analysing the traffic performance of these complex systems, a realistic model should take into account an estimate of the topological distribution of the failed satellites and the effects of their deterministic migration into the space network, resulting from the orbital dynamics.

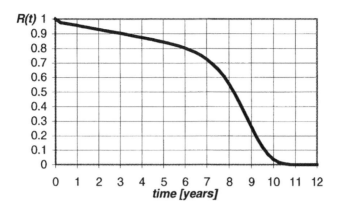

Figure 4.41   Reliability function for the operational and the spare satellite models

4.6 Security

4.6.1 Background

Interconnecting LANs through non-geostationary satellites, raises some specific problems related to security. Although many security issues encountered in LAN-interconnection via satellites are similar to those found in the Internet, there are situations when satellite-specific problems have to be addressed. However, the widespread popularity of the Internet Standards implies that the majority of solutions to those problems should adhere to such standards as much as possible. Whether security services will be provided through an ATM backbone network or through existing and new generation of IP networks, a security infrastructure will be required.

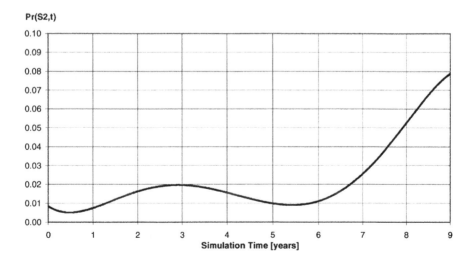

Figure 4.42   Probability Pr{Satellite Failed, t}

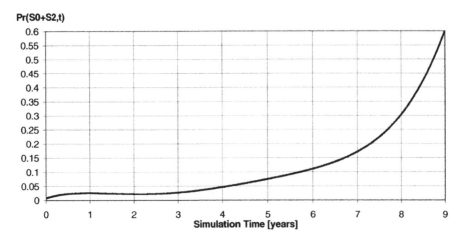

Figure 4.43   Probability Pr{No Service, t}

Extra security measures have to be imposed for satellite access to communication networks due to the following reasons:

- Eavesdropping and active intrusion is much easier than in terrestrial fixed or mobile networks because of the broadcast nature of satellites;
- Satellite channels experience high bit error rates which may cause a loss of security of synchronisation. This requires careful evaluation of encryption systems to prevent QoS (QoS) degradation due to security processing;
- Satellite networks, inherently, experience long delays, therefore satellite security systems must be efficient and should only add a minimum delay.

Security policies to counteract the above problems in satellite networks will have to be derived.

The organisation of this section is as follows: Before investigating the security infrastructure for LAN interconnection via satellites, a review on the progress of research in security is made in relation to Global System for Mobile Communications (GSM) security operation and UMTS security through the Advanced Security for Personal Communications Technologies (ASPeCT) projects. Following the review, security aspects of satellite ATM networks are outlined. Different proposals on the implementation of security mechanisms and service are studied.

4.6.2 Status of Current Research in Security in Communication Networks

4.6.2.1 Overview of GSM security

GSM provides security services (air link only) such as authentication, traffic confidentiality and key distribution. The communication link establishment is under the control of the Authentication Centre, which is often co-located with the Mobile Switching Centre (MSC). Every GSM subscriber has a Subscriber Identity Module (SIM) card containing a secret key known only to his Home Location Register (HLR) and the Authentication Centre. The mobile-station is authenticated in a signalling channel called the Dedicated Control

Channel (DCCH). If the authentication is successful, then the user is granted a Traffic (data) Channel (TCH).

When a GSM user switches on his/her mobile phone in a location other than the home location, the mobile station will notify the local MSC of its presence. The associated Visitor Location Register (VLR) will contact the user's HLR, which then requests a set of triplets from the Authentication Centre: a challenge (Random number RAND), a Signed Response (SRES) and the corresponding cipher (encryption) key. The set of triplets is then forwarded to the VLR. Each triplet is used only once for the authentication of the mobile station. The VLR send the challenge RAND to the mobile station. The mobile station will then reply with a SRES. Subsequently, the privacy between the mobile-station and the GSM network is achieved by enciphering data with the cipher key.

The main weakness of GSM security lies on the assumption that the terrestrial network infrastructure (VLR–HLR communications) is safe. This is not the case for global networks. In addition, GSM does not provide data integrity. It only provides unilateral user-to-network authentication. Network-to-user authentication is not provided.

However, the use of auxiliary signalling channels in GSM for authentication purposes can be conveniently adopted for satellite ATM networks.

4.6.2.2 Progress in UMTS Research in Security

The increasing demand for security by users, network operators and regulatory bodies, calls for more advanced security features in the third generation systems such as the Universal Mobile Telecommunications Systems (UMTS). UMTS security is studied in the ACTS project: Advanced Security for Personal Communications Technologies (ASPeCT – AC095). ASPeCT aims to specify advanced features based on public key security systems and propose solutions for migration from present GSM systems toward UMTS. Several proposals have been made by ASPeCT for User-network mutual authentication and air-link encryption [58]. However, there are several weaknesses in such proposals:

- It provides air link encryption only (i.e. User-Network similar to GSM) and does not permit end-to-end encryption between User-Service Provider;
- It does not allow security option negotiation such as encryption system and key lengths. This may cause inflexibility for the implementation of future security systems and incompatibility for services between countries with different security laws;
- It does not cater for ATM specific security issues which are important in satellite ATM networks.

However, the findings from ASPeCT can be used as a basis for deriving the security architecture of satellite ATM networks.

4.6.3 Security Services Implementation Issues

4.6.3.1 Encryption Algorithm

Two kinds of encryption algorithms are known: Symmetric algorithms and asymmetric algorithms. Symmetric algorithms are also known as secret-key algorithms. An example of such an algorithm is the Digital Encryption Standard (DES). Asymmetric algorithms are also know as public key algorithms, such as the RSAdigital signature system.

Symmetric algorithms use the same key for encryption and decryption whereas asymmetric algorithms use one key for encryption (e.g. private key) and another key for decryption (e.g. public key). The main disadvantage of symmetric algorithms is the risk of a key being revealed during transmission. Asymmetric algorithms were developed to eliminate such risk.

With asymmetric algorithms, there is no need to transmit the private key while the public key can be communicated with anyone, which reduces key-management complexity. It also provides a method for digital signature. However, the main disadvantage is their computational complexity.

The logical conclusion is to use both kinds of algorithms and their combinations to achieve optimal speed and security level. However, the issue of a properly authenticated public key is still a very critical issue in global networks.

4.6.3.2 Protocol Stacks for ATM Security

ATM technology serves as a basis for Broadband Integrated Services Digital Network (BISDN). It is vital to assure security services in such environment and when searching for security solutions, threats have to be identified first. A decision for security mechanisms and services along with their positioning within the ATM Protocol Reference Model (PRM) as shown in Figure 4.44, is made afterwards. In principle it is possible to place security services anywhere in the ATM PRM, however each placement has certain advantages and disadvantages. Specifically, the placement of encryption functions to the ATM protocol stack can be categorised into the following [63]:

- *Application level encryption*: The encryption function is located at the application layer, making it independent of the ATM. This has been proposed for special purpose applications, such as financial applications and stock trading, demanding application level security;
- *AAL level encryption*: AAL level encryption approach is alleged advantageous for LAN based traffic. The drawback of this positioning is that traffic flow confidentiality cannot be assured, as VCI/VPI are not encrypted. Furthermore, AAL usually operates as a single unit without accessible internal interfaces.
- *ATM cell level encryption*: The ATM layer adds five bytes for Virtual Path Identifier (VPI) and Virtual Channel Identifier (VCI) to the 48-byte Protocol Data Unit from the Segmentation and Reassembly (SAR) sub-layer. These five bytes are required to successfully route a cell to the final destination. Although maximum security can be achieved, placing encryption functions at this layer poses routing problems and requires encryption modules to operate at line speeds, which certainly is not a trivial task. Besides, as decryption has to take place at each switch to extract routing information, data becomes vulnerable. Key management will also become complex. If integrity is performed at this layer, additional segmentation functions would be needed.
- *AAL-ATM level encryption*: There is a clearly defined interface, which allows for a definition of a specific set of services (API). What should be taken into account is the fixed size of SAR PDUs, which should not be expanded when proper security mechanisms is used.

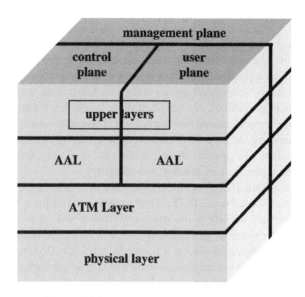

Figure 4.44   ATM protocol reference model

4.6.3.3 ATM Forum Security Specification

ATM Forum has defined the following four security services in the ATM security specifications version 1.0 [59] for terrestrial fixed networks:

- *User plane security*: The user plane security covers the Virtual Connection (VC) level, which can be either a Virtual Path Connection (VPC) or a Virtual Circuit Connection (VCC). It is subdivided into access control, authentication, data confidentiality and data integrity. It defines the mechanisms to allow for secure communication between nodes in an ATM network.
- *Control plane security*: The control plane defines the call control signalling functions required to establish, maintain and close a certain VC. Thus, authentication signalling has been defined as the main target of control plane security for any endpoint-to-endpoint, switch-to-switch, or endpoint-to-switch signalling communication.
- *Support services*: The support services define the certification infrastructures, the key exchange mechanisms and the basic negotiation of security requirements and capabilities.
- *Management plane security*: The management plane is responsible for both performing management functions for the system as a whole (plane management) and for performing network and system management functions such as resource management. The management plane security is not in the scope of the ATM Forum's preliminary security specification[80].

For data privacy, the ATM Forum has chosen the cell level approach by encrypting the ATM cell payload, whereas the ATM cell header is transmitted in the clear. For data integrity, ATM Forum has chosen AAL level and end-to-end integrity operating on the AAL-SDU.

4.6.4 IP Security–IPSEC

The most widely used protocol suite is TCP/IP due to the existence of a large number of quality applications that are written for this environment. It is therefore reasonable to assume that TCP/IP suite will be a predominant environment within broadband networks. However, current IP suite has many vulnerabilities including authentication threats (e.g. spoofing), confidentiality threats (e.g. sniffing), session overtaking (or highjacking) and integrity threats (an attacker intentionally tweaks some bits in packets). These threats are hard to prevent with the old TCP/IP technology, as they limit and complicate the use of large IP networks for sensitive communications. To prevent these threats, one needs cryptographic technologies that offer strong data authentication, privacy and integrity.

Problem of security within TCP/IP protocol suite have been intensively investigated within the Internet Engineering Task Force (IETF). IPv6 or IPSec [61,62] is a result of such investigation. IPSec products are already in the market. Within IPSEC, data authentication, privacy and integrity are provided at the network layer. A common security environment is established for all the applications within a particular network. Consequently a possibility is given to set-up the so called virtual private networks. These are logical networks that may span over many different physical networks, but are accessible only to users defined by a particular security policy. This means that one is no longer concerned with the physical protection of the network – any network provider with any technology can be chosen for LAN interconnections and the control of security remains within the LANs. The packet has a standard IP header, which can be routed with standard IP equipment and IPSEC therefore, provides backward compatibility with IP routers.

The basic ingredients of IPSEC protocol consists of the following:

- An authentication header (AH) for IP that assures authentication of the sender and the integrity of the data. The principle used is a verifiable signature;
- An encapsulation security payload (ESP) format for IP that encrypts data for confidentiality and supports many kinds of symmetric encryption (the default standard is DES);
- A negotiation protocol ISAKMP/Oakley to negotiate methods of secure communication and exchange of appropriate keys along with their usage.

Figure 4.45 shows the ESP format. The explanation of each field is as follows:

- The SPI is an arbitrary 32-bit number that specifies the device receiving the packet security related information such as algorithms, keys and their period of validity;
- The sequence number identifies the packets and indicates the number of packets that have been sent with the same group of parameters in order to prevent replay attacks;
- The payload contains the actual data information;
- The padding field is for encryption algorithms which requires integer multiples of bytes;
- The final field is for authentication.

In order to set-up a security association, the ISAKMP/Oakley makes use of two phases: the first phase is the establishment of a secure channel between peers; the second phase is the negotiation of general purpose Security Associations (SAs) between peers. There are three modes of exchanging keying information and setting up ISAKMP SAs. The first two modes, namely the Main mode and the Aggressive mode, are for the first phase and establish a secure

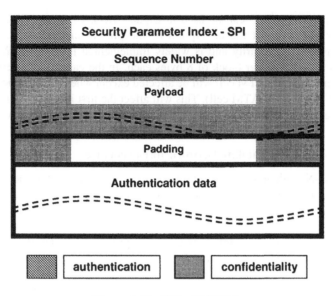

Figure 4.45   ESP of INSPEC

channel. The latter is simpler and faster. The third mode is the Quick mode which is for the second phase exchange by negotiating an SA for general purpose communication.

4.6.5 Security for ATM over Satellite

4.6.5.1 Transmission Rate and Encryption Key Updating

ATM has been designed for high data rates, therefore there is a need for a mechanism to frequently change the encryption key. The encryption block size in a typical cipher (such as DES) is 64 bits and no more than 10^9 blocks should use the same key to prevent intruders analysing the data [59]. The lifetime of an encryption key for ATM traffic at data rates of 155 Mbps is about 6.9 min and 8.9 h for data rates of 2 Mbps. Therefore at lower rates, a single encryption key can be used for the whole duration of connection while at higher data rates the key has to be changed frequently. The ATM Forum uses the master key concept to derive several session keys. Such a concept can be used for User-Service Provider (U-S) authentication and key exchange protocol in satellite ATM networks.

4.6.5.2 Encryption Synchronisation

Satellite channels normally experience high bit error rates and errors are of bursty nature. Therefore it is important to examine the impact of such errors on ATM cell payload encryption performance.

Two types of errors can occur. One is the complete loss of information bits (slips), which can be detected in the Transmission Convergence (TC) sub-layer through the cell delineation function before it reaches the ATM layer (and the encryption unit). Hence, this does not affect the encryption unit. The second type of error is the change of information bit values. (0-to-1 or 1-to-0). This type of error will affect the decryption outcome on the receiver.

For satellite networks, DES can be used in Cipher Block Chaining (CBC) mode where previous cipher blocks affect the encryption of the present cipher block. CBC mode is considered to have good security and is widely implemented [60]. An ATM payload can be divided into six DES cipher block of 8 bytes each. In the case of bit(s) errors in a DES cipher block, generally one ATM cell will be affected except when the error occurs in the 6th DES block (last block in the ATM cell). This will affect the first block of the next cell. In the worst case, bit errors in a single DES block will affect up to 1.1666 ATM cells.

4.6.5.3 Cost Reduction in Satellite Network Infrastructure

User-Network (U-N) authentication is important for the fraud prevention and revenue protection. However, information privacy and integrity is an end-to-end service and only concerns the User and the Service provider (S). Therefore, the role of satellite network operators can be confined to U-N mutual authentication only.

To reduce the cost of network infrastructure, data privacy and integrity data privacy and integrity can be omitted in satellite networks. End-to-end data encryption can be performed between U-S. To conform with ATM Forum specifications, encryption can be performed in the ATM cell payload only leaving a clear ATM header to allow for easy cell routing and switching by the satellite ATM network.

Digital signatures can be used to authenticate any other signalling and management messages exchanged between the user-network such as satellite handover and call termination messages.

4.7 Security Infrastructure

In order to provide any kind of security service on a large scale, a combination of symmetric and asymmetric algorithms has to be used. For authentication purposes, asymmetric algorithms will be used while for session encryption, symmetric ones will be deployed.

Mutual authentication protocol between the user and network operator conceals the user identity from intruders and implements digital signature and public key systems using asymmetric algorithms. Public keys can be stored in certificate servers which are also referred to certification authorities (CA). Every CA maintains a Certificate Revocation List (CRL) with invalid certificates. In this way, the whole problem of key management is now shifted to certificate management. To provide a globally operational security service, a security infrastructure to provide the following operational functions is required:

- CAs (or trusted third parties);
- A global electronic repository for distribution and maintenance of certificates and CRLs, like X.500 directory or WWW;
- Procedures for certificate management for issuing, distribution, prolongation, revocation.

However, problems remain to be solved in establishing such a security infrastructure. The basic problem was the rigid definition of X.509v1 certificate. Although X.509v2 includes two extra fields, the majority of problems remained. With the proposal of X.509v3 [ITU96], a solid ground for the establishment of security infrastructure can be formed.

When interconnecting LANs through non-geostationary satellites, the limitation in bandwidth on the air interface puts specific requirements on security infrastructure. It is likely that the basic X.509v3 format is too consuming for such purposes. More compact certificates with non-standard certificate format will have to be chosen. One such proposal is to simplify the role of CAs to the provision of secure billing services and end-to-end security services with key-escrow possibility.

There are many proposals on how to deal with authentication and key-exchange protocols in mobile LAN environments and possibly with their interconnection through non-geostationary satellites. It might be beneficial to start with crypto primitives which are inherently suitable for limited processing and bandwidth requirements. One of the most suitable cryptographic primitives for such purposes are strong one-way hash functions. These primitives can be used effectively to build authentication and key exchange protocols. These primitives have been studied intensively, especially in last few years, as they are frequently used for Mutual Authentication Codes (MACs). Such primitives and their protocols have been enhanced to overcome weaknesses discovered so far.

4.8 Conclusions

Some of the challenging problems encountered for the design of a broadband multimedia satellite network have been presented. The network architecture considered was a full-mesh constellation for the interconnection of LANs with multimedia services. The satellites are equipped with onboard processing such as switches and inter satellite link communication. These sophisticated functions require careful consideration of payload architecture, failure event possibilities and network management. The main issues related to system design have also been described.

Three case studies have been carried out to compare different routing strategies in non-geostationary satellite networks. The well-used approach of shortest path algorithm was applied together with the flow deviation of optimal routing techniques. Balanced and unbalanced traffic load and uniform and non-uniform distribution of Earth stations has been considered and some trials with the link length function have been investigated. Also the self-similar traffic model is examined versus classical Poisson model in order to simulate more realistic conditions of mode networks.

In any case, the performance of flow deviation techniques proved to be more reliable. However, due to the complicated topology of the system, a modification of the FD algorithm has been proposed and successfully applied, choosing only a limited number of paths to work with. A quantitative estimate of the number of paths was calculated through extended simulation running and it was proved that we can always find a very low number of paths to work without losing the quality of performance. The real-time technique for self-similar traffic illustrates that classic routing techniques, such as shortest path routing algorithms, result in a system degradation since the flow control mechanism of the network automatically decreases the throughput in order to avoid any congestion periods. On the other hand, optimal routing techniques succeed even in heavy traffic conditions to split the traffic among different paths (virtual circuits) and ensure the performance of the system. LEO constellations are very suitable for such routing algorithms because their network architecture presents symmetry and high degree of connectivity.

In addition, transparent procedures for hand-off, assuming zero bandwidth problems, have been considered, although this does not represent a realistic situation.

Comparisons have been made on two information distribution schemes for routing: the

Periodic scheme and the Triggered scheme. Both the connection blocking probability and the number of distributions are used to quantify the trade-off between route accuracy and signalling load. Results have shown that the triggered scheme prevents capacity consumption in unnecessary signalling of state information.

The concept of multicast was also analysed. Good support for group applications running on LANs interconnected by a satellite constellation requires that the satellites' on-board switches include native support for multicast. This appears unlikely in the commercially-proposed schemes discussed in this chapter.

Leaving implementation of multicast solely to the IP-routing ground networks, rather than making it a problem for both ground and mesh satellite networks would appear to make the problem of implementing efficient internetwork multicast with a satellite component more tractable.

SkyBridge can be considered the least ambitious of the proposed schemes when viewed from a satellite networking viewpoint, as the satellites are simple transponders rather than complex routing switches with intersatellite links. However, by being the most transparent from a tunnelling viewpoint SkyBridge paradoxically appears to offer the most, at present, for the implementation of IP multicast applications – at least until the appearance of native IP multicast routing on-board satellites.

The fixed position of geostationary satellites, including the Celestri GEO constellation, side-step the problems of implementing multicast support in a moving non-geostationary mesh and make the extension of existing legacy solutions, such as the Mbone, more practical.

The concept of reliability in satellite networks has been described. A reliability model has also been developed to guarantee a given level of availability for the space segment. The reliability functions for the operational and the spare satellite have been presented. Simulation has been carried out to study the probability of satellite failure and the probability of service unavailability.

Finally, current progress in security in GSM and UMTS has been reviewed. Security implementation issues in relation to satellite ATM networks have been discussed. Proposals have been made for interconnecting LANs through non-geostationary satellites using two PRMs: the ATM PRM and the TCP/IP PRM. The proposals are based on current trends and standardisation efforts, with consideration of specific needs for LANs with mobile devices that are interconnected through satellites. It has been suggested that security infrastructure which are used, should adopt policies implemented for Internet. Attention and possible solutions for specific needs in mobile environments, interconnected via non-stationary satellites have been addressed.

References

[1] COST226, 'Integrated Space/Terrestrial Networks', Final Symposium, Budapest, Hungary, 10–12 May, 1995.

[2] Z. Sun, T. Örs and B.G. Evans, 'ATM-over-Satellite demonstration of broadband network interconnection', *Computer Communications Journal*, Vol. 21, No. 12, pp. 1090–1101, 1998.

[3] T. Örs, T. Sammut, L. Wood and B.G. Evans, 'An Overview of Future Satellite Communication Options for LAN Interconnection', *3rd COST253 Management Committee Meeting TD(98)004*, Brussels, Belgium, 27th March 1998.

[4] 'Satellite ATM Networks: Architectures and Guidelines', *TIA/EIA Telecommunication Systems Bulletin 91 (TSB91)*, April 1998.

[5] ATM-Forum, 'Proposed Charter, Work Plan and Schedule for Wireless ATM Working Group', 96-0712/PLEN, 1996.

[6] T. Örs, Z. Sun and B.G. Evans, 'A MAC protocol for ATM over satellite', *Proceedings of Sixth IEE Conference on Telecommunications, IEE Conference Publication No. 451*, pp. 185–190, 1998.

[7] J. Bostic, T. Örs, M. Werner, H. Bischl and B.G. Evans, 'Multiple Access Protocols for ATM Over Low Earth Orbit Satellites', 1st Joint COST252/259 Workshop, pp. 77–87, 1998.

[8] ITUT-T Study Group 13, 'B-ISDN semi-permanent connection availability', *Draft new Recommendation I.357*, Geneva, May 1996.

[9] B.G. Evans, *Satellite Communication Systems*, 2nd Edition, London, Peregrinus, 1991.

[10] M. Annoni, G. Garofalo and C. Ratti, 'Dynamic Resources Assignment Strategies for Advanced Communication Systems', INTERCOM'93, Vancouver BC, Canada, 22–25, 1993

[11] M. Annoni, 'Reliability Modelling for Satellite Constellation Systems', TD(99)-6, February 1999.

[12] N. Celandroni, E. Ferro, F. Potorti and G. Maral, 'Delay analysis for interlan traffic using two suitable TDMA satellite access schemes', *International Journal of Satellite Communications*, Vol. 15, No. 4, pp. 170–175, 1997.

[13] H. Koraitim, 'Multiple Access Protocols and Resource Allocation Over a Satellite Channel', PhD dissertation, Ecole Nationale Superieure des Telecommunications, ENST 98 E 014, September 1998.

[14] H. Koraitim and S. Tohmé, COST253 TD(98) 020: 'Resource allocation and connection admission control in satellite networks'.

[15] H. Koraitim and S. Tohmé, 'A Movable-Boundary Integrated CBR/Bursty-Data Traffic in Star-ConFig.d VSAT Satellite Networks', *IEEE Symposium on Computers and Communications ISCC'97*, Alexandria, Egypt, June 1997.

[16] H. Koraitim and S. Tohmé, 'Movable Boundary Policies for Resource Allocation in Satellite Links', *ICCC'97 Conference*, Cannes, France, November 1997.

[17] H. Koraitim and S. Tohmé, 'The Impact of the Threshold Value On the Performance of the DMBS Allocation Scheme', *Networld and Interop98 Engineers Conference*, Las Vegas, NV, May 1998.

[18] H. Koraitim, S. Tohmé, M. Berrada and A. Brajal, 'Performance of Multiple Access Protocols in Geo-Stationary Satellite Systems', *IFIP HPN'98 Conference*, September 98, Vienna, Austria.

[19] H. Koraitim and S. Tohmé, 'Resource allocation and connection admission control in satellite networks', *IEEE Journal of Selected Areas in Communication*, Vol. 17, No. 2, 1999.

[20] H. Koraitim and S. Tohmé, 'A Multiple Access Protocol for Packet Satellite Networks', *ISCC'99*, Sharm El Sheikh, Egypt.

[21] H.S. Chang, B.W. Kim, C.G. Lee, Y.H. Choi, S.L. Min, H.S. Yang, D.N. Kim and C.S. Kim, 'Performance comparison of static routing and dynamic routing in low-earth orbit satellite networks', *Proceedings of the VTC'96*, pp. 1240–1243, 1996.

[22] I. El Khamlichi and L. Franck, 'Study of two policies for implementing routing algorithms in satellite constellations', *Proceedings of the 18th AIAA ISCSS'2000*, 2000.

[23] W. Krewel, G. Maral, 'Analysis of the impact of handover strategies on the QoS of satellite diversity based communication systems', *Proceedings of the 18th AIAA ISCSS'2000*, 2000.

[24] E. Papapetrou, et al., 'Performance evaluation of LEO satellite constellations with inter-satellite links under self-similar and Poisson traffic', *International Journal of Satellite Communications*, Vol. 17, pp. 51–64, 1999.

[25] R.A. Raines, R. Janoso, D. Gallagher and D. Coulliette, 'Simulation of two routing protocols operating in a low Earth orbit satellite network environment', *Proceedings of the 1997 IEEE Military Communication Conference*, Vol. 1, pp. 429–433, 1997.

[26] H. Uzunalioglu, 'Probabilistic routing protocol for low Earth orbit satellite networks', *Proceedings of IEEE ICC'98*, June 1998.

[27] G. Berndl, M. Werner and B. Edmaier, 'Performance of Optimized Routing in LEO Intersatellite Link Networks', *Proceedings of IEEE 47th Vehicular Technology Conference*, Vol. 1, pp. 246–250, May 1997.

[28] A. Shaikh, et al., 'Dynamics of quality-of-service routing with inaccurate link-state information', *AT&T Research Report, 1997*.

[29] M. Steenstrup, *Routing in Communication Networks*, 1st edition, Eaglewood Cliffs, NJ, Prentice-Hall.

[30] M. Werner and G. Maral, 'Traffic flows and dynamic routing in LEO intersatellite link networks', *Proceedings of IMSC'97*, Pasadena, CA, pp. 283–288, 1997.

[31] M. Werner, C. Delucchi, H.J. Vogel, G. Maral and J.J. De Ridder, 'ATM based routing in LEO/MEO satellite networks with intersatellite links', *IEEE Journal of Selected Areas in Communication*, Vol. 15, No. 1, 1997.

[32] M. Werner, 'A dynamic routing concept for ATM based satellite personal communication networks', *IEEE Journal of Selected Areas in Communication*, Vol. 15, No. 8, 1997.

[33] A.H. Ballard, 'Rosette constellations of Earth satellite', *IEEE Transactions on Aerospace and Electronic Systems*, Vol. AES-16, No. 5, pp. 656–673, 1980.

[34] E. Papapetrou, I. Gragopoulos and F.N. Pavlidou, 'Performance evaluation of LEO satellite constellations with inter-satellite links under self-similar and Poisson traffic', *Internation Journal of Satellite Communication*, Vol. 17, pp. 51–64, 1999.

[35] D. Bertsekas and L. Callager, Data Networks, 2nd Edition, Eaglewood Cliffs, NJ, Prentice-Hall, 1992.

[36] E.W. Dijkstra, 'A note on two problems in connection with graphs' *Numerische Mathematik*, Vol. 1, pp. 269–271, 1959.

[37] M.A. Sturza, 'Architecture of the Teledesic satellite system', *Proceedings of the International Mobile Satellite Conference*, pp. 212–218, 1995

[38] P.A. Worfolk and R.E. Thurman, 'SaVi – software for the visualization and analysis of satellite constellations', originally developed at The Geometry Center, University of Minnesota.

[39] L. Wood, 'Network performance of non-geostationary constellations equipped with intersatellite links', MSc thesis, University of Surrey, November 1995.

[40] P.L. Spector, J.H. Olson, et al., FCC filing of February 28, 1997, 'Application of SkyBridge LLC for authority to launch and operate the SkyBridge system ', SkyBridge LLC.

[41] M.D. Kennedy, B. Lambergman, et al., FCC filing of June 1997, 'Application to construct, launch and operate the Celestri multimedia LEO system', Motorola Global Communications, Inc.

[42] S. Deering, 'Host extensions for IP multicasting', RFC1112, August 1989.

[43] D. Waitzman, C. Partridge, S. Deering, et al., 'Distance vector multicast routing protocol', RFC1075, November 1988.

[44] J. Moy, 'OSPF Version 2', RFC2178, July 1997.

[45] S. Deering, D. Estrin, D. Farinacci, M. Handley, A. Helmy, V. Jacobson, C. Liu, P. Sharma, D. Thaler and L. Wei, 'Protocol Independent Multicast – Sparse Mode (PIM–SM): Protocol Specification', RFC 2117, June 1997.

[46] D. Estrin, D. Farinacci, et al., 'Protocol Independent Multicast – Sparse Mode (PIM–SM): Protocol Specification', RFC 2118, June 1997.

[47] A. Bestavros and G. Kim, 'TCP Boston: a fragmentation-tolerant TCP protocol for ATM networks', *Proceedings of Infocom '97*, 1997.

[48] M. Annoni, 'Reliability Modelling for Satellite Constellation Systems', COST253 TD(99)-006.

[49] M. Annoni and S. Bizzarri, 'An integrated modelling methodology for the simulation of the life cycle of satellite constellations', Joint COST252/253/255 Workshop, Toulouse, 19–20 May 1999.

[50] P.D.T. O'Connor, *Practical Reliability Engineering*, 2nd Edition, New York, Wiley, 1985.

[51] A. Villemeur, *Reliability Availability, Maintainability and Safety Assessment*, Vol. 1: Methods and Techniques', New York, Wiley, 1992.

[52] NASA/LeRC OS&MA, 'Reliability Block Diagram Tutorial', on-line NASA / Lewis Research Center – Office of Safety & Mission Assurance Documents.

[43] M. Annoni, 'ITN-260 – RBDAN (Reliability Block Diagram Analyzer) – Program Design Specification', IRIDIUM Technical Note, Motorola Satcom.

[54] R.E. Barlow and F. Proschan, *Statistical Theory of Reliability and Life Testing. International Series in Decision Processes*, New York, Holt, Rinehart and Winston, 1975.

[55] M. Annoni and S. Bizzarri, 'A simulation environment for the evaluation of maintainability strategies in telecommunication networks based on satellite constellation', *2nd European Workshop on Mobile/Personal Satcoms (EMTS'96)*, Rome, Italy, 9–11 October 1996, London, Springer-Verlag.

[56] M. Annoni, S. Bizzarri and F. Faggi, 'CONSIM: A flexible approach to satellite constellation simulation', *DSP'98. 6th International Workshop on Digital Signal Processing Techniques for Space Application*, 23–25 September 1998.

[57] M. Annoni, S. Bizzarri and F. Faggi, 'Performance evaluation of satellite constellations, the CONSIM simulator concept and architecture', *3rd European Workshop on Mobile/Personal Satcoms (EMPS'98)*, Venezia Lido, Italy, 4–5 November, 1998, London, Springer-Verlag.

[58] 'Advanced Security for Personal Communication Technologies – Migration/Evolution towards UMTS Security issues', ASPeCT Project.

[59] ATM Forum Security Specification Version 1.0 draft 1998, ATM Forum.

[60] B. Schneier, *Applied Cryptography*, New York, Wiley, 1996.

[61] K.S., Atkinson, 'IP Authentication Header', RFC 2402.

[62] K.S.,Atkinson, 'IP Encapsulating Security Payload', RFC 2406.

[63] V. Vardharajan, R. Shankaran and M. Hitchens, et al., 'A Practical Approach to Design and Management of Secure ATM Networks', *20th National Information Security System Conference*, Baltimore, MD, 1997.

5

Evaluation Tools

Complex communication systems involving satellite links are difficult to test and tune-up without the help of simulation tools. First of all, the use of satellites is very expensive; the time spent in testing and tuning-up the system must be as short as possible. Secondly, during the performance evaluation in a real environment it is not always possible to find the right amount of traffic, and the most appropriate traffic pattern and data aggregation that will put the system under the maximum amount of stress so that its limits can be validated. These problems are worse when satellite constellations, rather than single satellites, are involved, and in all but unique cases, it is virtually impossible for the researcher to test real satellite constellations.

It is therefore necessary to use some kind of simulation tool. The array of required features for the tool is very broad. The basic feature is making it possible to describe the constellation topology, and the position of Earth terminals. Satellites may have different capabilities in terms of number and characteristics of intersatellite links, spot topology and behaviour, and on-board processing abilities. Many types of routing algorithms are under study, some of which allow for dynamic re-routing based on the nodes' perceived congestion or on Quality of Service (QoS) requests. It should also be possible to analyse channel allocation algorithms for the up-down links, including frequency reuse algorithms. The traffic generation choice should be broad and easily extensible, allowing for both fine control of single connections and generation of background traffic for loading the satellite network. Finally, a rich set of statistic collection and analysis tools should be available.

Existing simulators are generally bulky and expensive. Moreover, most of them are general-purpose tools, and are not very well-suited to the particular environment of Low Earth Orbit (LEO) and Medium Earth Orbit (MEO) satellite constellations [17]. Free software is a desired feature for a new simulator for satellite constellations. This would allow any researcher to enhance the source of the simulator. If the original architecture of the simulator is well-done, the simulator will grow in features and reliability, and will eventually become a tool for research co-operation between different research institutions, both academic and commercial.

In order to reach this goal, the simulator should be modular, so that new features can be added without changing the overall structure. Also, the use of an easily portable programming language would be an advantage. In order to make it possible to tackle complex simulations, the simulator should allow parallel processing of concurrent activities. After a general overview of network simulators presented in Section 5.1, Section 5.2 presents LeoSim, a simulator for studying routing algorithms in LEO environments. LeoSim operates at the connection level. Section 5.3 covers GaliLEO, an ambitious project aimed at the provision of a free, extensible simulator for the study of access protocols, resource allocation and routing algorithms in constellation environments. GaliLEO is the result of a joint effort from some participants of the COST253 Action. Section 5.4 addresses the simulation of failures in constellation through the presentation of CONSIM™ (CONstellation SIMulator). CONSIM™ can be interfaced with GaliLEO. A tool dedicated to the study of routing algorithms, AristotleLEO, is described in Section 5.5. Finally, Section 5.6 provides an overview of SEESAWS, an IST proposal issued by some organisations involved in COST253. SEESAWS goal is to provide a general purpose environment for evaluating the performance of satellite networks. It encompasses the aspects covered by the simulators presented in prior sections by using an hybrid simulation/emulation approach.

5.1 An Overview of Network Simulators

Most of the simulation tools which can be found in the literature are *discrete-event* simulators, i.e. they model a certain system as it evolves over time by representing which state variables change at a countable number of points in time. These points in time are the ones at which an event occurs, where an event is defined as an instantaneous occurrence which may change the state of the system. Let us examine some of the most popular tools.

OPNET is a simulation tool for analysing communication networks by using models [1]. It is based on an extended finite state machine and is written in C. OPNET models are specified in terms of objects, each with configurable sets of attributes. These attributes can be specified either by a graphical process, an associated text file, or as variable parameters in the simulation description file. The specification of the model is organised into a hierarchy of four different levels: network; node; process; and parameter. When the model is specified, OPNET generates a simulation program written in C. Although C is not a particularly suitable language for modelling and simulation, the choice of UNIX/C guarantees the portability of the system.

BONeS is a simulation system for studying communication network models [2]. BONeS SatLab is a flexible, special software package for the design, animated visualisation and analysis of satellite-based communication systems. SatLab can be used to model mobile/satellite systems, which may include satellites, fixed Earth stations, and moving vehicles/persons. It provides multiple animated views of global satellite systems, uplink, crosslink and downlink analysis, jamming, and adjacent satellite interference analysis representation of fixed, mobile and portable Earth stations. Three types of simulation can be performed: positioning; design; and communication simulation. The positioning simulation allows the user model to analyse and animate various configurations of satellites, fixed Earth stations and moving stations. The design simulation automatically performs simulation for a range of parameter values. For example, the user can determine the interference of communication satellite systems on other communication systems or the probability of interference between satellite systems for selected frequencies. The communication simulation is used to track the source and route of data packets and to determine their best route based on relative distance, velocity, angle, visibility, traffic congestion, and interference between two or more nodes. Unfortunately, there is a strong uncertainty about the future of BONeS, and currently this is a strong deterrent in using it.

RESQ is a software package developed at IBM Research for defining and solving extended queuing network models [3,4]. Problems which have been analytically solved typically fall into the class of queuing systems for modelling the performance of computer communication systems. For the analytical solution of a queuing model, assumptions must be made about the system modelled. Typically these assumptions relate to the process of arrivals at the queues, the service processes at the queues, and the scheduling disciplines. RESQ provides a numerical solution component, QNET4, which uses the convolution algorithm for product form networks, and a simulation component, named APLOMB. RESQ is especially strong in the statistical analysis of simulation outputs and the determination of appropriate simulation run lengths.

AMS is an environment which integrates facilities and tools to build a communication system, to study its performance, and to validate the connectivity [5]. The user of the 'atelier' can construct a concise system in a graphical environment and execute it, starting from models of several standard networks, such as LANs (Ethernet, Token Ring, Fibre Distributed Digital Interface (FDDI)), WANs (X25, Transmission Control Protocol/Internet Protocol (TCP/IP)), satellite (Time Division Multiple Access (TDMA), FDMA) and radio-networks available in a specific library. AMS was designed on the basis of existing and proven packages, such as QNAP2, GSS (Graphical Support System), MODLINE and S-PLUS.

Other simulators can be cited, such as Network Simulator (NS) (http://www.mach.cs.berkeley.edu/ns) or Ptolemy (ptolemy.cecs.berkeley.edu), but their characteristics are not adequate for satellite environments. For example, NS is TCP oriented, and only a limited support to satellites is now operational; while Ptolemy is too general and is not network oriented. Nevertheless, Section 5.3.6 investigates how NS could fit in a framework such as GaliLeo.

5.2 LeoSim: A Simulator for Routing

Today's communication networks are complex. When the need for a performance evaluation arises, analytical studies are often not tractable considering the number of interacting mechanisms as well as their complexity. Of course, it is possible to render such studies tractable by oversimplifying the behavioural model of the mechanisms or by studying them independently. However, the resulting performance evaluation may then not reflect reality. Considering the economical stake involved in satellite constellation deployment, performance results should be as realistic as possible. These considerations call for simulation-based studies.

Simulations are now limited by the power of today's computers. Simulating satellite networks suffer inherently from this problem due to the wide-coverage of the network and the broadband type of the offered services. Therefore, given these constraints, the design of simulation tools should optimise the processing power used. One way to cope with this power limitation problem is to design modular or component-based simulators. Such simulators consist of components interacting together, each with clearly defined tasks. Complex models are implemented in components that are crucial with respect to the study while simpler models are implemented in the others components. It is then possible to tightly control the processing power needed by each component. Furthermore, component-based simulators are easier to distribute over a network in order to share the processing load among many computers.

It is also possible to design components which, instead of providing well-defined behaviours, specify rules that describe the component properties. These components are called generic components and cannot be directly used to provide simulation results. They serve instead as templates to create concrete components which will have behaviours fulfilling the rules specified by the generic components. This technique makes it easier to re-use or customise simulators.

LeoSim uses this technique and is more accurately described as a generic routing simulator for satellite constellations [6]. LeoSim is a simulation framework in which components (such as routing algorithms, resources managers, etc.) are plugged in and evaluated. In order to design such a framework, it is essential to establish a network model.

5.2.1 Network Model as Supported by LeoSim

LeoSim is intended to study QoS routing in satellite constellations with inter-satellite links. Therefore, the network model implemented is built around routing and focuses on aspects relevant to the study of routing performance.

The network model supported by LeoSim is based on the following assumptions. The constellation provides services to fixed or mobile users. When mobile users are considered, the assumption is made that, due to the high satellites velocity, the users can be treated as fixed. Since LeoSim is intended to study QoS routing, the services use a connection-oriented communication mode. This decision is motivated by the fact that QoS is difficult to implement in connection-less environments. As a matter of fact, both ATM and IntServ use connection oriented mode. Satellites are equipped with an on-board processing capability. This on-board capability includes routing and resource management.

Routing in satellite constellations is split into two processes: up/down-link (UDL) routing and inter-satellite link (ISL) routing. These two processes ask for different algorithms and therefore are modelled as two different components in the simulator. The UDL Routing component selects the start and end satellites taking into account the source and destination Earth stations, as well as the connection requirements. The ISL routing component computes routes between the start and end satellites, taking into account the connection requirements.

In order to make the appropriate decisions, the routing algorithms must be provided with information about the network resources status. The scope of such information can include the information concerning the entire network and send it to adjacent satellites. In any case, this information has to be collected and distributed by each satellite, and this is the purpose of the component called the *topology manager*.

The information distributed by the topology manager presents a summary of the status of the resources located in the satellites and on the different links. These resources are modelled in LeoSim as three components called satellite resources, ISL resources and UDL resources. They are referred to as network resources and are allocated to the connections in the satellite network. Connections express their requirements in terms of user QoS. There is a need for a component in each satellite to be able to map user QoS into network resources that are allocated to the connection. This component is called the *QoS manager* since it is responsible for ensuring the agreed QoS between the user and the network. The QoS manager provides additional services such as allocation/release of resources and call admission control.

A call signalling protocol supervises the different operations, such as network resources allocation and release, and this protocol must be realised in order to establish and to release a connection. LeoSim incorporates a call signalling component to perform call signalling.

The dynamics of the connection requests are modelled by another component of LeoSim called the *call generator*. The call generators are attached to traffic sources and show the characteristics of newly-created connections in terms of destination station, traffic requirements, connection duration and connection inter-arrival time. LeoSim also supports the generation of traffic confined in the constellation through the use of special call generators called *constellation restricted call generators*. By using such call generators, it is possible to load the network with background traffic using a less detailed, thus less processing power-demanding, traffic model.

It must be noted that, for the moment, LeoSim supports a component for call generation but no component for packet generation. Packet level simulation in broadband networks raises performance issues that are far from trivial. However, in medium term future, LeoSim will support packet level simulation. It will then be possible to achieve a compromise by having most of the traffic supported on connections without packets and a residual part of the traffic with packets. Packet level simulation enables the study of time related properties such as end-to-end delay or jitter.

Since LeoSim applies to satellite constellations, there are situations where a connection is broken due to a handover. A component called the *handover manager* deals with such situations and performs, for example, re-routing.

Finally, the constellation animator is a component implementing the orbit mechanics and providing services such as satellite location, ISL and UDL length computation, ISL and UDL holding times evaluation, UDL elevation, etc.

Figure 5.1 gives an overview of the different components that are present in LeoSim, as

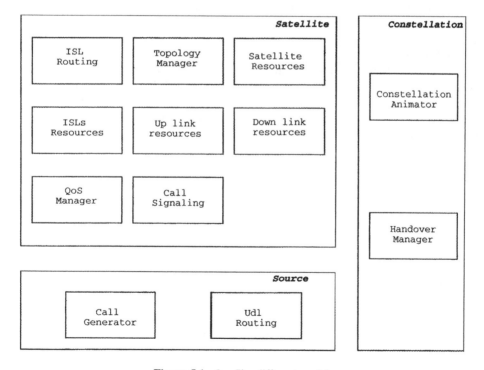

Figure 5.1   LeoSim different models

well as their physical locations in a real system. The handover manager could have been located in the satellite. However, since LeoSim is not aimed at the study of handover signalling, it makes sense to have only one handover manager for all satellites.

Until now, LeoSim was presented as a satellite network simulator. It can also be used to study terrestrial wired networks as well. Indeed, considering the sources as end systems (following the Open System Interconnection (OSI) terminology) and satellites as intermediate systems, the satellite network can be mapped into a wired network. Some components used in this context are more simple because the intermediate systems are not mobile and handover does not occur.

5.2.2 Introducing Versatility in the Network Model

As stated earlier, LeoSim is more a simulation framework than a complete, self contained simulator. Some of the components presented in the previous section are generic, defining only templates based on which concrete components will be developed later. This section exemplifies the concept of generic component. Examples of studies that can be undertaken using LeoSim and a short description of the concrete components already developed are provided.

5.2.2.1 Generic Components Supported by LeoSim

In order to leave LeoSim as open as possible to changes, all components except the call signalling are generic. Instead of having well-defined behaviours coded in the components, the generic components define a set of rules that implementations must fulfil. As an example, the ISL routing component is generic. It specifies a rule stating that in order to build a concrete ISL routing component, one must provide a procedure to compute a route given a connection description (in terms of source, destination and requirements). The actual algorithm used to compute the route is not described in the generic ISL routing component but it is specified later when a concrete component will be derived from the generic component. Figure 5.2 illustrates how two concrete ISL routing components are built, based on the generic ISL component.

5.2.2.2 Example of Studies Possible with LeoSim

Table 5.1 shows examples of studies that can be realised with LeoSim. These studies are associated in the table with the generic component to which they relate. For example, the study of handover techniques requires concrete handover managers derived from the generic handover manager and implementing the handover management techniques evaluated. Having a non-generic handover manager does not provide the opportunity to allow such a study. Indeed, the handover manager technique would be frozen in the simulator and could not be changed without modifying the simulator code and possibly the simulator design.

5.2.2.3 The LeoSim Library of Concrete Components

Some concrete components have been developed in order to test the validity of LeoSim framework and concepts. These are:

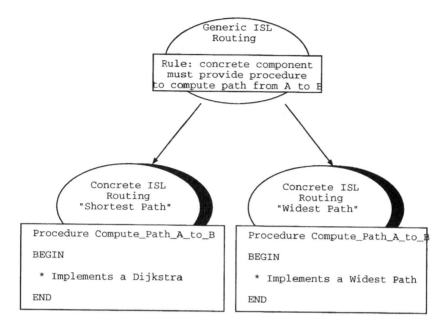

Figure 5.2   ISL routing components built based on generic ISL component

- A QoS manager which allocates and releases by subtracting and adding the reserved capacity from the available capacity. The call admission control verifies that enough capacity is present in the equipment according to the connection requirements expressed in terms of capacity.
- An ISL routing algorithm that minimises the route length in terms of hops. The routing algorithm takes into account the capacity requirement of the connection.
- Three topology managers implementing three operating modes. The first mode broadcasts information periodically. The second one broadcasts information when the resources status in the satellite has changed by a given amount. The third one broadcasts information only once, at the beginning of the simulation.

Table 5.1   Example of possible studies using LeoSim

Generic component	Study
UDL and ISL routing	Study of candidate metrics and multi-criteria optimisation for routing
Topology manager	Evaluation of the traffic load caused by routing information distribution
Handover manager	Study of handover techniques such as partial re-routing
QoS manager and network resources	Evaluation of routing performance using different resources management schemes
Call generator	Evaluation of routing performance under different traffic loads

- An UDL routing algorithm that selects the start and end satellites according to the longest visibility time.
- A constellation animator for polar type constellations (such as Iridium) and inclined constellations (such as Globalstar).
- A handover manager performing complete re-routing for connections broken due to hand-overs.
- Several call generators implementing different traffic models.

The simulation performance measurements are expressed in terms of call blocking probability, number of re-routings and other specific metrics. Additional performance metrics can be added, however it requires modification to the simulator core program. Such modifications are not necessary when adding concrete components. In the next section we will present the different implementation techniques that were used to create the simulation framework.

5.2.3 Implementation Issues

We will first present the technique used to implement the generic components. Then we will cover all aspects relevant to the way components collaborate. Finally, we will present some performance issues that arise as a consequence of the implementation choices that were made.

5.2.3.1 Supporting Generic Components

In order to implement the generic capability of LeoSim we needed a programming language with the following features:

- The ability to group data and processing aspects of a component into a single structure.
- The ability to specify the services that concrete components have to implement in the generic components.
- The ability to add concrete components without modifying the simulator core and with a minimum of programming effort, by re-using existing code.

All these considerations led us to use an object-oriented language, namely Java [8–11,16]. Java was preferred over C++ because we consider it cleaner and safer. As an example we perceive garbage collection as a must when dealing with concrete programming. Finally, we did not choose Eiffel, Camel or YAFL because Java has a wider support. The performance issues related to Java will be covered later.

With an object-oriented language, it is easy to implement the generic capability of LeoSim. Abstract classes are used to define generic components. When a concrete component is added, Java has to extend (or derive from) the abstract class defining the equivalent generic component. Using abstract classes makes it possible to ensure that concrete components comply with the rules specified in the generic component.

Finally, Java features the ability to load new classes in the system during run-time. This capability is used during the configuration phase of the LeoSim. When the configuration file is parsed, the simulator core loads all classes related to the concrete components used for the simulation. Compliance of the concrete components with the generic components is tested during this phase.

Having presented how the components are integrated individually in the LeoSim framework, we will now cover how they inter-operate.

5.2.3.2 Making Components Inter-Operate

Component inter-operation covers two aspects:

1. The way components use services provided by other components.
2. The way checks are made ensuring that concrete components are able to collaborate together.

The first point is solved by implementing a discrete event simulator [18]. When the simulator is started, all needed concrete components are loaded. A component invokes a service of another component (called the target component) by posting an event to the scheduler. The event specifies the target component as well as the service that must be invoked. Additional information such as time before invocation and service parameters are included. The scheduler performs some checks in order to verify whether the target component offers the requested service.

We will illustrate with an example the issue raised in the second point. A concrete QoS manager knows only about some concrete network resources. A mechanism must exist which, during the configuration of LeoSim, checks that concrete components are compatible with each others. For this reason, each QoS manager must provide a method of illustrating, by its name, whether it is able to handle network resources. The same technique is used in other components.

To summarise, during the configuration phase of the simulator the following tasks are undertaken:

- All classes related to the concrete components used in the simulation are loaded in the simulator.
- When appropriate, concrete components are checked in order to verify that they know how to co-operate with the other components.

By using the techniques presented above, it has been possible to develop a flexible, although robust, tool. We believe that adding or customising components is not a complex task. Nevertheless, we will be able to determine the actual learning curve when distributing LeoSim to other users.

5.2.3.3 Performance Issues

These issues can be categorised as follows:

1. Issues related to the nature of broadband network simulations.
2. Issues related to the use of Java, which is interpreted in its native form.

The first point is related to the number of events generated during the simulation. Even if the processing power needed to treat a single event is low – handling millions of events can be a rough task for a medium size computer. Our objective is to minimise the number of events generated. This fact motivated the decision that LeoSim is, up to now, a call level simulator rather than a packet level simulator. We also simplified the call signalling implementation.

Generally speaking, events correspond to network packets whatever the type of packet: data or control packet. Implementing a realistic call signalling consists in having events sent for each signalling packet transmitted among network equipment. Considering the number of connections set-up during a simulation run (several tens of thousand), this can result in a fairly high number of events. In our implementation of the call signalling, the start satellite directly invokes the appropriate methods in the other satellites to handle call set-up and call tear down. This bypasses the event scheduler, thus reducing the number of events, but preventing from detailed investigations about signalling delay.

When designing concrete components, the user of LeoSim has to be aware that the more events are generated, the more time it will take to run the simulation. Method invocation should be used as often as possible instead of event passing.

Finally, the LeoSim event passing mechanism was designed in such a way that it eases the transition to and from event passing and method call, without causing modifications to the code. This facility is useful in order to tune the performance.

Java, in its native form, is an interpreted language. The Java source code is compiled into an intermediate language called the *bytecode*. The program is then executed by a virtual machine. The interpretation results in a performance decrease of about 15% compared to native machine code execution. Nevertheless, tools exist or are emerging, that compile Java into machine code. The 1.2 release of the Java Developer Kit includes such a tool. It is also worth mentioning the Cygnus project aiming at building a Java front-end for gcc. As a consequence, we do not consider that the choice of Java introduces any performance penalty.

Java supports two different paradigms for parallel computing. The first one, called threading, is oriented towards multi-processor computers. The second one, called remote method invocation, offers a convenient mechanism to build distributed applications sharing their processing load over a network.

At present time, LeoSim only makes use of threading. Two event schedulers are available. The first one is dedicated to single processor computers. The second one can take advantage of SMP computers by assigning the event processing to different processors. An additional mechanism is implemented to ensure that the parallel processing does not result in causal relationships violation.

5.3 GaliLEO: A Framework for Joint Expertise

GaliLEO is a simulator for constellations of communication satellites. GaliLEO is work in progress. Its purpose is to simulate the transmission of both connection oriented and connection-less traffic over a constellation of LEO/MEO satellites. The aim is to evaluate the performance of various constellation access techniques, resource management schemes, and routing policies [13,14]. GaliLEO, is written in Java and will interface with the CONSIM™ simulator by CSELT [7], which is designed to study the impact of faults on system performance.

GaliLEO should eventually become a free software simulator available to the scientific community, providing common tool for studies on satellite communications. An overview of some characteristics of GaliLEO is presented, followed by details on the current status of the implementation and directions for future development.

GaliLEO is a comprehensive simulator for satellite constellations, targeted towards LEO/MEO communication systems. GaliLEO is work in progress: its aim is to provide a testbed for studying various aspects of communications using satellite constellations. In the satellite access arena, it will be used for developing and evaluating satellite channel access techniques and beam frequency allocation algorithms. As far as routing is concerned, we are planning studies relative to UDL routing, ISL routing, and integrated routing; all of these possibly with QoS constraints. The simulator should be able to handle a realistic number of connections going on.

The initial design of the GaliLEO architecture was the outcome of a collaboration between two institutes where researchers had already had experiences with developing simulators. GaliLEO is the result of separate experiences carried out on two different simulators, LeoSim and SimToc. These two tools do not include all the features we needed. Other tools on the market are either used to simulate only some specific aspects of the transmissions on the LEO satellite constellations, or are simply too costly to maintain. Hence the creation of GaliLEO, which aims to be a general purpose, customisable tool, freely available to the academic world.

We have gone from the design stage to the beginning of implementation, thanks also to the help from three students working at CNUCE, in Pisa (I). Previous simulation projects have created the background for GaliLEO: they include SimToc and LeoSim, developed at CNUCE/CNR (I) and ENST (F), respectively; and FRACAS (FRAmed Channel Access Simulator) developed at CNUCE. GaliLEO is thought to be a synthesis of the concepts born with previous works, powerful enough to provide a common test bed for the academic community of researchers in the satellite communications area.

LeoSim, one of GaliLEO's ancestors, is an event-driven, continuous time simulator written in Java and developed at ENST in order to study one specific topic: link state routing algorithms in LEO satellite constellations. LeoSim relies on object-oriented techniques in order to be easily adapted to various routing algorithms. It provides statistics on the number of call requests, the call block probability, and the transmission overhead introduced by maintaining the link state database. LeoSim is continuously evolving; its current version supports several constellations and ISL routing algorithms. By means of a user graphical interface it is possible to specify various parameters of the constellation and routing algorithms, and to follow the real-time evolution of the simulation. It is possible to dynamically load modules implementing different behaviours of the constellations, routing, and the handover management. Its basic design approach, as well as the event dispatcher engine, have been transported into GaliLEO. SimToc, the other ancestor, was focused on the station and the UDL between the station and a satellite of the constellation; its architecture is no longer being developed.

Another simulator, CONSIM™, which was developed at CSELT (I) for evaluating the performance of satellite constellations affected by partial failures, has been integrated 'by results' into GaliLEO. Integration by results means that the two simulators are kept separate, and the results of one program are fed as input to the other one. This is probably the best way to accommodate two simulators that were written separately, and to minimise the coupling needed between the different teams responsible for the programs. This approach is generally feasible only when the two studies cover aspects which can be independently studied. Our approach is to use CONSIM™ in order to produce a list of faults which, according to the model, will occur during the constellation's lifetime. Each of these failure events specifies the type of failure as well as the time of occurrence. This list will then be used to feed GaliLEO's simulator engine in order to trigger the right fault managers at the appropriate time.

The language used for developing GaliLEO is Java. Although this language is yet immature, at least as far as performance is concerned, it has many interesting characteristics. In fact, it provides a well-designed object-oriented multi-platform environment; it is well-backed by industry and academy; and is easier to use than the common OO standard, that is, C++ [12]. We hope to see significant improvements on the performance side of Java in the near future, so we may end up having a finite product by the time the performance of Java platforms is comparable to that of a natively compiled C++ program. All libraries used by GaliLEO are free, basing the foundations for a wide diffusion and collective improvement of the product.

5.3.1 Both Global and Limited Coverage

When studying traffic pattern distributions, or routing algorithms, or congestion patterns depending on the time of the day, it is useful to model an entire satellite network having global terrestrial coverage (Figure 5.3). GaliLEO is able to emulate a complete constellation, and run simulations involving Earth stations distributed on the whole Earth surface. The movement of satellites is tracked and their relative positions are computed.

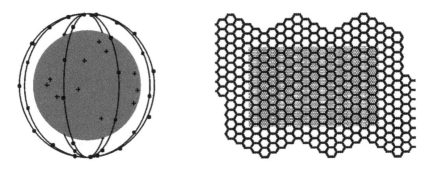

Figure 5.3   The model of the Earth

When the simulation concentrates on network access policies, or beam frequency allocation and reallocation, then a geographically limited simulation is sufficient. In this case, GaliLEO is able to track the behaviour of the Earth stations and satellites in the limited geographical region, while simulating the outside environment using a given statistic model.

The possibility of simulating only a limited region relieves the user from the burden of specifying detailed station positions and traffic patterns on the whole globe when the study is focused on the area covered by one or a few satellites only (e.g. when complex frequency reuse strategies are considered, or when the effect of a complex traffic pattern over a particular channel access policy is studied). The simulation is also faster when compared with a simulation that involves a detailed description of the whole globe.

5.3.2 The Architecture

The main components of the GaliLEO architecture are shown in Figure 5.4.

'Input' consists of the input files containing the simulation run specifications (i.e. station location, traffic patterns, channel allocation policies, constellation characteristics, type of statistics to collect, parameters of the simulation engine). 'Validator' performs the syntactic

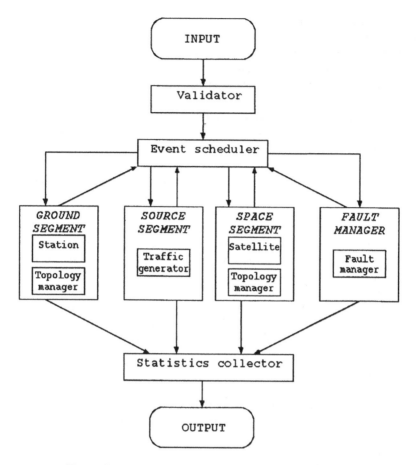

Figure 5.4   Main components of the GaliLEO architecture

and semantic validation of the input data. 'Event scheduler' analyses the queue of events and executes the actions triggered by the scheduled events. The 'topology manager' in the 'ground segment' and 'space segment' keeps track of respectively the 'station' and the 'satellite' movements. The 'fault manager' manages the failures which may happen to any of the elements of the simulated system. The 'traffic generator' implements the traffic model. The 'statistics collector' collects data from the simulated model, computes the required statistics and dumps data for subsequent off-line analysis.

5.3.3 Assumptions and Definitions

Here are some terms and definitions of important concepts and entities defined inside GaliLEO.

A *cell* is an area of the Earth illuminated by one satellite spot beam. A *footprint* is the whole coverage area of a satellite, i.e. the union of the areas covered by its spot beams. An *overlap area* is the area in which a station (i.e. a single subscriber or a concentrator) can receive a signal with an acceptable power level from more than one adjacent spot beam. An UDL is the aggregation of all spot beams pertaining to the same footprint; it has a fixed capacity, and is uni-directional. A *beam* is the communication medium between a satellite and a spot on the ground. A beam has a variable capacity which must not exceed the capacity of the UDL to which the beam belongs. A node of the network is any station or satellite. Satellites have multibeam antennas for uplink reception and downlink transmission, and are connected to neighbouring satellites by means of *inter-satellite links* (ISL) which are uni-directional. A link is either an inter-satellite link or an up-down link. A *generator* is any producer of communications; regular (or ground) generators communicate between a number of pairs of stations, while constellation generators communicate between two satellites: the latter are used for testing, debugging, or generating background traffic during simulations.

A handover (or hand-off) occurs when either a UDL connecting a satellite to a station is cut-off, or when a beam change occurs (inside the same UDL), or when an ISL is cut off. All connections passing through that link must be re-routed. Connections are assumed to be full-duplex, with forward and return channels, where forward channels are intended to be from source to destination, and return channels from destination to source. A call connection drop occurs when an existing connection has to be dropped. It may happen either when there is a handover and the connection cannot be re-routed, or when high priority traffic pre-empts all the resources used by a connection. Analogously, a partial drop occurs when a connection must be reduced due to one of the above reasons. A call block occurs when a new connection cannot be established. It may happen when there are no resources available in the network in order to support the new connection.

A satellite is associated with a *space position* which varies deterministically over time. Satellite movements are described through orbital mechanics; a station is associated with a *ground position* which may vary randomly or deterministically over time if the station is mobile.

Traffic generators can generate both call connections and data. Data can be transmitted over connections (connection oriented traffic) or in best-effort mode (connection-less traffic). In the following, when *data* is used, we refer to both connection oriented and connection-less data, even though the simulation of connection-less data transmission over a LEO constellation will be not implemented during the first stage of this project. In fact, implementing a complete simulation of a complex connection-less network is currently an open issue, due to the fact that the computing power required should be comparable to that of all the nodes involved in the real network, which is next to impossible. However, in the future, limited scope simulations of connection-less traffic may be considered, so the architecture of Gali-LEO will provide hooks for that kind of simulation.

5.3.4 Logical Behaviour

Figure 5.5 shows the logical diagram of GaliLEO's behaviour. The source module is the dispatcher of all the actions related to call generators. During the initialisation phase, it asks all the generators what will be the time of their next action. It then schedules an event for each generator. When an event is triggered, the source performs the relevant action, informs the relevant generator of the event, and asks when it is time to perform the next action, then schedules an event for that time. Connection set-up and connection modifications are not

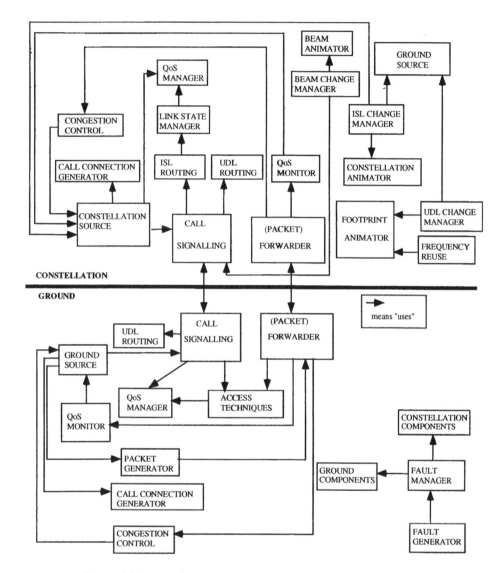

Figure 5.5 Logical diagram of the architectural core of the simulator

notified immediately to the source upon completions. Instead, the source module sets up a timer, that generates an event which simulates the delay necessary for the operation. Only when the event is triggered, is the source informed of the outcome of the action and asked for the time of next action.

The source performs all these actions for both types of connection generators, ground generators or constellation generators. The set of operations is exactly the same, only the delay for set-up connections is always set to null, and the constellation generator never asks for modifications of the connection.

A call generator object provides a model of the behaviour of connection sources. It provides a set of entry points for creating a connection object, calculates times for the next connection modification and computes the time for the next connection set-up. A connection modification may be either a connection release or a change in the connection's parameters, such as requested QoS or size. A generator which changes the size of a single connection can emulate a concentrator of phone calls with a variable number of terrestrial connections multiplexed onto a single satellite connection, or a data connection whose size varies its bandwidth on demand. A single generator may handle an arbitrary number of connections. This may be useful for simulating traffic which is correlated between different pairs of stations, such as multicast traffic which goes over unicast connections.

The *packet* generator module implements the traffic models for the packets running over a connection.

The *UDL routing* module selects the first and the last satellite where traffic enters and exits the constellation. In order to choose the satellites, both the geometrical configuration of the satellites and the traffic load are considered.

The access techniques module deals with connection-less traffic and handles the assignment of slots within a MAC frame. The module is provided with information about the traffic (e.g. load and traffic type) and computes the bandwidth assignments accordingly.

The footprint animator module tracks the state of the UDLs between all satellites and all stations. In other words, the footprint animator is responsible for saying the couple satellite-station in communication.

The call signalling module is responsible for establishing, releasing and modifying the connections. All these operations involve interactions with the node resource manager.

The forwarder module provides services to switch the traffic packets. The switching is done either by using a connection identifier (for connection oriented communications) or by using default routes (for connection-less communications).

The *QoS manager* module is responsible for the management of the equipment resources as well as the translation from one requirement/resource description to another. The QoS manager also performs call admission control for a given satellite or station. The *QoS monitor* module ensures that packets relevant to a given connection are compliant with the connection QoS contract. The *congestion control* module monitors the resource status in the station or in the satellite as a result of connection-less traffic. If congestion occurs, appropriate actions are taken such as signalling the sources or dropping packets.

The *frequency reuse manager* module takes care of frequency allocation for each spot beam of a satellite. The *ISL change manager* module handles all changes in ISL characteristics: this may happen in connection with re-routing if, for example, an ISL is switched off. The *UDL change manager* module plays the same role as the ISL change manager for up and downlinks. The *fault manager* module implements the fault reaction model. Notifications of faults are generated by the fault generator (this entity is an interface with results provided by CONSIM™) and sent to the appropriate satellite components.

The *ISL routing module* computes the connection route within the constellation, between the entry point satellite and the exit point satellite as provided by the UDL routing module. If the ISL routing algorithms used are of link state type, a link state map must be maintained at each node, which reflects the state of the whole network topology. In order to construct the link state map, each node broadcasts information about its status and gathers information broadcast by other nodes.

The *link state manager* module handles all the broadcasting and gathering of information

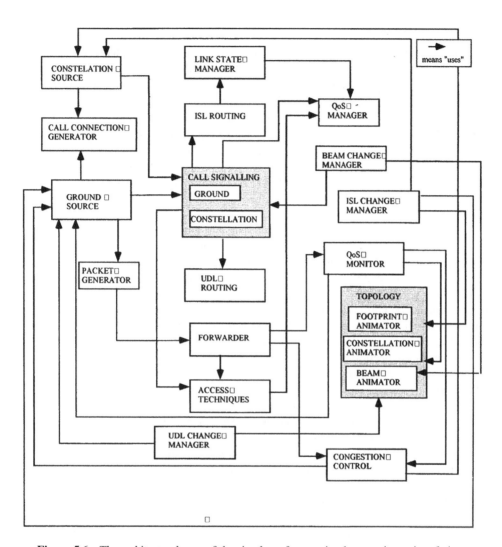

Figure 5.6   The architectural core of the simulator from an implementation point of view

and maintains a link state map. The *topology manager* module implements the orbital mechanics and therefore the satellite movements. It detects the occurrence of changes in length occurring in the ISLs during satellite movement and whether an ISL is switched on or off. The ISL properties are updated according to the satellite positions. The *beam animator* module tracks the state of a single beam within a given footprint. It detects the occurrence of changes in length in the beams during satellite movement as well as the beam moving from spot to spot. The beam change manager module reflects the changes in beam configuration, on the satellite, as detected by the beam animator. The *beam change manager* invokes, for example, re-routing if the result of the beam change is such that a connection can no longer be supported on that beam.

Figure 5.6 shows the above modules organised from an implementation point of view.

5.3.5 The Connection Set-up

The number of channels of a connection is not fixed after a connection has been set-up, but can change during the lifetime of the connection. This can simulate connections coming from a call concentrator (*aggregated phone calls*); e.g. a concentrator may set-up a single connection for all the phone calls it handles, and may simulate both new phone calls and old closed phone calls by varying the number of channels used by the single connection set-up at the start time. In other words, a number of n phone calls from station i to station j is simulated by the generation, in station i, of a unique connection that requests n channels. The model of a connection is more general than this, allowing the dynamic variation of all the requested characteristics of a connection. For example, the model can be used to emulate a data connection whose requested QoS and size vary, depending on the link quality perceived by the application and the user requests.

The connection set-up model is synchronous, so that routines called in sequence allocate the resources (according to their availability) in a simple way. This makes it impossible to analyse the connection set-up time in detail, as this model makes it instantaneous. However, it is possible to approximate the connection set-up and modification times: once a connection has been set-up or modified, the generator controlling it is not immediately notified of the change; instead, the notification is delayed by a given amount of time. This amount of time could also be made dependent on the nodes that the connection is crossing, thus making it more adhering to the simulated network model. In any case, the resources on the nodes traversed by the connection are instantaneously occupied, and the delay only involves notifying the controlling generator.

The connection set-up procedure is triggered by the source module which asks the controlling generator to create a connection and then passes it as an argument to the set-up routine of the source station. The routine returns a connection whose requirements are less than or equal to the one originally created by the generator, depending on the resources available on the route from the source station to the destination station. If enough resources are available, the returned connection is the same as the one passed as an argument to the set-up routine; otherwise it can be reduced in size or QoS. The connection request may also get completely blocked due to a lack of resources. The set-up is present in every node – that is, in each station and each satellite. When a set-up routine in a given node is called with a connection as its argument, the node first checks for local availability of resources. If the connection can only be partially accommodated because of limited resources, the number of forward/return channels in the connection is reduced accordingly, and the QoS possibly changed. The connection is then passed as an argument to the set-up routine of the next node towards the destination. When the set-up routine of the next node returns the possibly reduced connection, local resources in the node are then allocated for the returned connection, i.e. the state of the node is updated, and the connection is returned to the caller.

At this stage resources have been allocated in all the nodes traversed by the connection. The connection is made of two distinct unidirectional connections, which share the source and destination stations, but may have otherwise different characteristics. The forward and backward paths of a given connection may have different sizes, QoS characteristics, and may be routed through different sets of nodes. They may have to satisfy certain relative constraints. For example, when a phone call concentrator makes a request, it must set-up equal-sized forward and backward unidirectional connections. Such requirements imply that the connection may be further modified after all the resources have been allocated on the traversed nodes. This is the responsibility of the call signalling module of the source station.

The call connections are variable in size. Connections may be reduced in size or QoS characteristics either as a consequence of a handover, or because of the pre-emption of resources (for example, when high priority traffic needs part of all of the resources used by lower priority connections).

When the time comes for the set-up or modification of a connection, an event is triggered which calls the source module. The source module asks the call generator to create a new connection object or to modify the characteristics of an already existing connection object, in case of modification. The source module then calls the source station call signalling module, which performs the operations that are described above. The call signalling module of the source station returns a possibly modified connection object to the source modules. This connection is the one that has been installed in the forward and backward routes, i.e. the connection for which resources have been occupied in the relevant network nodes. At this point, the source generates a time delay after which it notifies the controlling call generator of the outcome of the requested set-up or modification operation. The source module also asks the controlling call generator for the time the next modification on that connection will happen in the future, and generates an event for that time, which will call the source module connection modification routine.

Here is a list of some limitations of the connection model implemented in GaliLEO.

- Only point-to-point connections are considered: multicast or broadcast connections are not supported and can be modelled using a number of unicast connections.
- A connection cannot be split on more than one path, e.g. all the phone calls managed by a concentrator must follow the same route, however forward and return channels are not necessarily on the same path.
- No re-routing is performed as a consequence of modifying a connection; this means that a request for higher bandwidth may fail, even if the requested bandwidth is available on a different path from the source to the destination.
- It is possible to emulate a delay for the call signalling of connection set-up and modification, however the resources on the nodes traversed by the connection are instantaneously allocated. This is only an approximation of the behaviour of a real network, where the resources are allocated at different times depending on the propagation delays of the call signalling.

5.3.6 Traffic Over a Connection

While we foresee a time when GaliLEO will be able to generate packets to be injected inside a connection, the first stages of its implementation will not provide such a facility. However, it is possible to obtain most of its functionality by using a two-stage approach. Let us suppose one sets up a complex simulation environment using a version of GaliLEO which is capable of considering connections with varying characteristics, such as delay, bandwidth and QoS (bit error rate, for example). After running a set of simulations, one may be able to obtain a set of traces describing the time varying nature of a connection between two given endpoints. These traces can then be used as input data for a different simulator, which uses them to model a link's behaviour where it injects packets. Such an arrangement would be a useful way of analysing, for example, the impact of varying bandwidth and bit error rate over TCP connections traversing a LEO constellation. The simulator to use for this kind of study could be the Network Simulator (NS), which is freely available [25,26].

GaliLEO would provide the basis for generating the background traffic, for managing all the details of the mobile network and the routing, for taking care of call block, drop and partial drop events, possibly caused by higher-priority traffic in a pre-emptive allocation algorithm. NS would take care of all the details relative to the protocol stack implementation, and to all the variations on TCP behaviour that have been developed for use with NS.

The limitations of the described approach stem from its two-stage nature, which does not allow any sort of feedback between the first and second steps. For example, with this approach it is not possible to model an application that changes its QoS or bandwidth requirements depending on its perceived link quality. In order to simulate such an application, the two stages need to be integrated, that is, the packet generation and protocol stack behaviour must be included in GaliLEO.

5.3.7 The Topology

The *topology module* is responsible for computing the absolute and relative position of all the geographical objects in the simulator, i.e. the stations and the satellites. It is also responsible for computing the times at which various events happen. It is composed of three sub-modules: the *footprint animator*, the *constellation animator* and the *beam animator*. The functions of the topology module can be split into three domains: the *constellation topology*, defined in terms of satellites and ISLs, the *ground topology*, defined in terms of stations, and the *UDL topology* (Figure 5.7).

All these topologies are dynamic because satellites move with respect to both the stations and to each other; ISLs are switched on and off due to pointing, acquisition, and tracking requirements; up and downlinks are also switched on and off because of visibility requirements; stations may be mobile.

The space reference system for the topology modules may be continuous or discrete. The

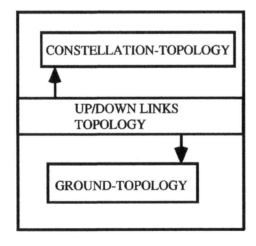

Figure 5.7   The topology module

most natural reference system as far as the satellite positions are concerned is continuous, so the constellation topology module uses a continuous space reference, expressed as latitude, longitude and altitude of each satellite. In some cases a discrete space reference can be useful for the ground topology. For example, a constellation like Teledesic, which implements a satellite spot-hopping capability, may be simpler to describe using a latch of cells for describing the Earth surface geographical reference system. The UDL topology could make use of the space references chosen for the space and ground topologies, however we prefer a discrete space reference in station-satellite units. Note that the time-space relationship involves both satellite movements and the Earth's spin. For most constellations the satellites are at the same altitude, so a more compact space co-ordinate form may be used, which only involves latitude and longitude values.

5.3.7.1 The Constellation Topology

This module contains routines for computing the satellite positions. These are used for computing the link length and status. In fact, links can be activated or deactivated depending on satellite visibility and other parameters, such as the visibility angle, the satellite latitude, the speed of change of the length or the speed of change of the visibility angle. When a simulator entity (generally the ISL Change Manager module) wants to be notified of a link state change, it calls a monitoring routine inside the topology module. At the time at which the interesting event occurs is computed, an event is generated. All the entities that made a call to the monitor for a given link state change are put into a queue related to that link. Once the event is triggered, all the interested entities are notified and removed from the queue. Interesting events include the activity switch of a link, which can be on or off, and the change of a link's length by more than a given threshold, which can be used for evaluating the QoS characteristics of a link, such as delay and bit error rate.

The ISL change manager module is responsible for checking that the connections using the modified ISL are still valid. The connections may need re-routing because of changed QoS parameters depending on the link length, or because the link is finishing.

The module is initialised by reading an input file where the constellation parameters are specified. The first constellation type that is being implemented is one where the orbit planes are evenly distributed apart from the ascension angle at the seam, and they all have the same declination angle. The orbits all have the same number of satellites, the same altitude, and the satellites in the orbit are equally spaced. The phase of the satellites with respect to their orbit varies by a fixed amount from one orbit to the next one. The parameters needed for defining this constellation type are:

- Number of orbital planes;
- Declination of orbital planes;
- Ascension angle at the seam;
- Orbit altitude;
- Number of satellites per orbit;
- Phase between the corresponding satellites of each orbit.

Such a model represents most orbits of practical interest. However, any kind of constellation model can be implemented in the above described constellation topology framework, which is extremely general.

Other parameters needed for a communication constellation are the number of inter-orbit and intra-orbit ISLs per satellite, and a condition modelling for switching them on and off. The simplest conditions are visibility and satellite latitude, but provisions for more sophisticated conditions are made, should the need arise to consider them.

During the initialisation phase, the constellation topology is responsible for creating all the necessary satellite and link objects and initialising them, assigning a status and a position to each. After that, computations are performed only on request, i.e. when some entity subscribes to the monitor in order to receive notification of a given event. This architecture makes it conceptually possible to reduce the computations to a minimum.

5.3.7.2 The Ground Topology

The ground topology has entry points which return the time-dependent position of a station by using a co-ordinate system which depends on the chosen space reference. It also computes the position of a station with respect to another station and the list of nearby stations; this information is used for frequency reuse computation.

An input file contains the position of each station. For mobile stations, the input file also describes the type and parameters of the movement.

5.3.7.3 The UDL Topology

The UDL topology covers both the dynamics of the footprints (footprint animator module) and the dynamics of the beams within a given footprint (module). It is accessed through monitors, in a way similar to that used for the constellation topology module. The events that can be monitored are the station changes with respect to a given footprint. Specifically, it is possible to be notified when a station enters or exits a footprint. State changes of beams can also be monitored, i.e. when a beam goes on or off due to length or visibility angle thresholds exceeded, and when the length of a beam changes by more than a given threshold, analogously to ISLs.

The UDL topology module relies on the information contained in the satellite and station databases that are managed by the constellation and ground topology modules, respectively. During the initialisation phase, this module reads the information pertaining to the geometry of the spots inside a satellite's footprint from an input file, and initialises the relevant data structures in the satellites.

5.3.8 Routing

In the LEO constellation context, routing is usually split into UDL routing and inter satellite links (ISL) routing. In the GaliLEO model, Uplink (UL) routing is the process by which the source station selects the source satellite used to forward the packets of the connection, while Downlink (DL) routing is the process by which the destination station selects the destination satellite from which the packets of the connection will arrive. Possible criteria used for UDL routing are the availability of resources at the satellite and at the station, the minimisation of the handover rate on the UDL, and the quality of the communication between the station and the satellite.

Given a source satellite and a destination satellite, chosen as the result of UDL routing, ISL routing computes the optimal path between these two satellites. Possible criteria used are: resource availability in the satellites and ISLs; minimisation of the handover rate; quality of the communication among satellites; and length of the path.

Route computation is performed on the basis of information which describes the network state. The network state is defined as the buffer occupancy in the nodes and the geometry of the constellation. This type of information must be made available to the nodes which perform the route computations. A module, called *link state manager*, is dedicated to the distribution/gathering of such network state related information. By implementing different policies in the routing algorithm and in the link state manager, it is possible to model centralised/decentralised/distributed algorithms. GaliLEO framework relies heavily on the generic (using inheritance and polymorphism). The routing algorithm, the link state manager and the description of the different resources can be tailored according to the study objectives.

5.3.9 Fault Management

Today's satellites are getting more and more complex in order to support the current broadband services. Increasing the payload complexity leads to an increase in the number of possible fault sources and also a financial impact. Consequently it is crucial to accurately model the reliability of a satellite constellation. By *reliability* we mean the probability that a given failure does not occur in a certain time frame. Reliability can thus be completely described with a probability density function. The goal of reliability modelling is therefore to select the proper density function for each element to be included in a model and to calculate the resulting reliability accordingly. Before trying to compute the density function, it is mandatory to give a precise description of the system (i.e. its architecture) for which we are trying to define a reliability model. Some assumptions have to be made. An obvious example is the use of ISL. It is relevant to the present analysis, since ISLs need dedicated hardware which must be considered in the reliability model. The various faults that might occur can be roughly categorised into *partial* and *catastrophic* failures. While the former still allow the satellite to work, even if in a degraded mode, the latter make the satellite completely stop working. Each failure then has to be weighted over time using a probability density function. The choice of these density functions can be complicated and necessitates an extensive understanding of how a satellite is made up. Finally, all these density functions (or their respective reliability functions) must be merged in order to compute the general reliability function of the satellite. Note that even if two failures are only partial, this can still be fatal to the satellite. Therefore the general reliability must cope with the reliability of each element as well as the relationships among different failures. Trying to compute this reliability function is thus quite complex and many simulations have to be used in order to have an estimate of the reliability function. GaliLEO will not compute the reliability function by itself, since CONSIM™ is dedicated to this. Rather, the simulation engine of GaliLEO will be fed with events notifying failures. The nature and time distribution of these events is provided by CONSIM™.

5.3.10 Some Implementation Aspects

This section covers issues related to the implementation and performance of GaliLEO. In broadband network simulation, performance is a critical issue. As a matter of fact, the throughput of the equipment and links, combined with the global coverage of the satellite network makes simulation of such architectures a challenge. An improvement of the simulation speed gives the opportunity to simulate more traffic or to add complexity to the simulation model. Optimising simulation is therefore not only a matter of reducing the time required for a simulation run but also to push the limits further. Various optimisation techniques may be used at different levels of the simulator: at the model level; the design level, and the implementation level.

At the model level, exhaustive simulation of given mechanisms can be replaced by behavioural models, provided they are less costly to simulate. The increase in simulation speed has a cost, however, since the behavioural models often provides an approximation of the reality. In broadband network simulation, a critical issue is how to model the effect of the packets on the queuing delays. Simulation of packets is costly, given the involved data rates. Usual techniques consists of deriving results from the queuing theory to estimate the average waiting time of a customer (a packet in this case) in the system. However, the results obtained from queuing theory assume that the arrival of packets follow given laws which might not be suitable to our purposes.

At the design level, the issue of whether to use parallel or distributed simulation arises. Parallel or distributed simulation can be implemented using multiprocessor computer or cluster of computers connected by means of a low latency network. Distributed simulation increases significantly the complexity of the simulator, mainly because of the requirement to ensure causal relationship. This goal can be achieved in two ways. Conservative systems introduce synchronisation among the distributed nodes, decreasing the performance of the overall system. For this reason, another class of systems (called optimistic) were introduced. In optimistic systems a curative approach is implemented where no synchronisation is used until causal relationship is violated. On such occasions, the system is rolled back to the last known safe state. Supporting roll back might a tedious (and memory consuming) task, however in situations where roll back is seldom needed, optimistic systems are more efficient.

In order to support distributed simulation, a mechanism must exist for exchanging data among nodes and invoking pieces of code located remotely. GaliLEO is written in Java, the remote method invocation is therefore a candidate to provide such facilities. GaliLEO does not support distributed simulation [15] from the beginning but it is scheduled within a medium time frame. The direction currently foreseen is a cluster of PCs (i.e. distributed simulation) rather than a multi-processor platform (i.e. parallel simulation). On the one hand, multi-processor systems benefit of fast Central Processing Unit (CPU) interconnection but the scalability is still limited to few CPUs. On the other hand, entry class multi-processor systems (typically bi-processor PCs) are interesting for relieving the CPU of some tasks such as the management of the graphical user interface and input/output (I/O) operations. Java supports threading (also known as lightweight processes) and intensively uses threading for its Graphic User Interface (GUI) components. LeoSim uses a separate thread for the output of results on file. Therefore, a temporary blocking of the I/O operation does not affect the simulation course.

At the implementation level it is possible to reduce the execution time of the simulations by using optimisation techniques and appropriate tools. Java was first aimed at the development of easily deployable Internet applications. Current Java compilers do not produce native (assembly) code but rather an intermediate language called bytecode. The bytecode is then interpreted by means of a Java Virtual Machine (JVM). The interpretation of bytecode does not match the speed of native C/C++ code but for most applications, the performance is still acceptable. Obviously it is not the case for simulations where the faster is the better. A first solution is to use JVMs that translate, the bytecode into native code. An alternative is to use native compilers that are now appearing on the market and produce native code from a Java source.

Apart from using appropriate tools, the traditional profiling/optimising techniques can be applied. Special attention has to be paid to the memory-related issues due to the memory management scheme used in Java. C and C++ are often blamed for they need to free previously allocated memory blocks. Programs become prone to memory leaks and other heap corruption bugs. On the other hand, Java does not require such explicit freeing procedure. Rather, a process known as the garbage collection keeps a watchful eye on allocated memory blocks and claim them back when they are no more used and when the system is getting low on memory. Programming in a language supporting garbage collection is definitely safer and more comfortable. However, garbage collection can severely impair the system performance if a lot of objects are allocated and then discarded. Garbage collection is performed frequently and will take considerable time. The key to efficiency in this context is to favour object re-usability (or recycling) rather than throwing away the object for later garbage collection.

Finally, the need for efficient data structures cannot be emphasised sufficiently. For example, saving some clock cycles in the handling of a connection table is worthwhile considering that a typical simulation run accounts for millions of connections that are set-up and then released.

5.3.11 GaliLEO Methodology and Project Management

GaliLEO's project life cycle is following a spiral approach based on a core simulator which is gradually being enhanced. The analysis and design are object-oriented since the usage of object orientation is one of the corner stones of GaliLEO's general nature. The methodology relies heavily on diagrams which follow the UML standard as well as some internal guidelines. All diagrams are accessible in HTML, from a Web server. The same applies for the source code, which is accessible from a Web CVS repository. Currently, the primary development and analysis platform is Linux and commercial software is not being used. We hope to be able to carry on like this.

Work has begun on implementation of parts of GaliLEO, with the help of three computer science students. Each student worked on a separate piece of the whole picture. More work on the GaliLEO project is going to begin. The arguments of the three theses are described in the next three sub-sections.

5.3.11.1 An Interactive Graphic Interface for Defining Constellations

This work produced a standalone application useful as a tool for creating input data for GaliLEO. It was also the base for a possible future module embedded inside GaliLEO. This Java application has a graphic interface consisting of one or more forms for entering constellation data and a graphic window showing a picture of the revolving Earth, the satellite orbits as lines, and the satellite positions as dots moving along the lines. Using the Java3D library guarantees portability and continued support of new features and performance enhancements.

The main purpose is the quick prototyping of different constellations, and the error-free generation of input data for GaliLEO. This application is able to produce a configuration file for GaliLEO, describing the constellation parameters, and text tabular data with the detailed satellite positions, useful for precise checking and tuning of the constellation.

Advanced features include the ability of creating stations on Earth by clicking on the Earth positions – and accordingly generating a station configuration file for GaliLEO – and the ability of being connected with GaliLEO's topology module in order to show the ISL and UDL links as they are set-up and torn down.

5.3.11.2 Topology Class

This is a complete implementation of the routines (methods in Java terminology) for computing various parameters related to the satellite positions.

Examples regarding the constellation part include computing a given satellite position at a given time, the relative positions of two satellites, the time when a given link will be set-up or torn down, the length of a link, the time when the link length will change by a given quantum.

Examples regarding the UDL part include the time when an UDL link will be set-up or torn down, the list of satellites usable by a station at a given time, the length and elevation of a given UDL link, the time when the length or elevation of a given UDL length will change by a given quantum.

Examples regarding the station position include the position of each station; the positions are currently static, but provisions are made for moving stations simulating mobile users or concentrators, such as ships or planes.

The topology class also implements functions for keeping track of beam configurations and beam frequency allocations.

5.3.11.3 Generator Class

This thesis was centred on a partial implementation of the Generator class, including only one generator type for point-to-point connections with exponentially distributed connection time lengths, handling multiple connections simultaneously, with Poisson distribution of the number of connections [19]. Advanced features for this generator include using a time-of-day dependent matrix of traffic rates between multiple stations.

In order to make the generator work, it was necessary to implement a complete interface with the rest of GaliLEO, and to write stub implementations of many other important classes, thus providing a sound starting point for a first working version of GaliLEO. In particular, the Connection, Link, Station, and Satellite classes were implemented, and connection block, set-up, drop, tear down functions were tested with basic functionality, thus allowing testing of the main flow paths in GaliLEO's logic.

5.4 CONSIM™: A Complementary Tool for Reliability

The problem of controlling and maintaining a constellation of satellites by considering the effect of the occurrence of partial or catastrophic failures is quite unexplored and constitutes one of the main issues covered by CONSIM™. The CONSIM™ simulator prototype has been identified and is used to analyse the impact on the service provided by examining the occurrence of partial failures in the satellites of a constellation or to analyse the performance and effectiveness of a given maintainability policy. The purpose of these analyses is to understand if a given measure should be taken; which one, among a number of alternatives, should be chosen; when to take a measure; and what is the estimated cost of the considered maintenance policy [20–24].

The general architecture of the CONSIM™ framework and its interface with the external environment are represented in Figure 5.8. In order to simplify the inter-working with any external software tool (i.e. off-line simulations or analysis, model customisation and input parameter definition, statistical or graphical data post-processing, etc.), the interface is based

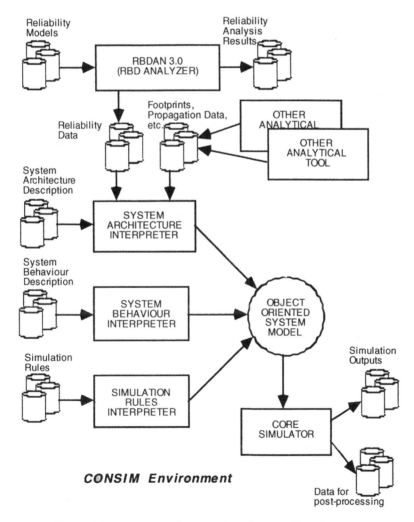

Figure 5.8 General architecture of the CONSIM™ environment

on a set of ASCII files easy to edit, process and display with any computer, independently of used CPU or Operating System.

Basically, the environment requires the CONSIM™ user to provide, by means of ASCII files, a description of the satellite system to be simulated in terms of physical architecture and logical behaviour.

Additional data resulting from off-line external tools (i.e. reliability analysis, antenna footprints, propagation statistics, etc.) can be imported in the CONSIM™ environment. As an example, Figure 5.8 illustrates RBDAN, a predictive reliability modelling and analysis tool developed in CSELT. Finally, the user must also identify the statistical parameters to be analysed during the simulation.

The complete set of user-provided data is used by CONSIM™ to specialise the embedded general-purpose model template and to automatically create all the software entities needed to properly simulate the system. A syntactic and semantics control on the consistency and completeness of the data provided by the user has been introduced because of the initial operation, which is essential for the correct working of the simulator, and the large number of parameters supplied by the user. This control is performed by means of three dedicated 'interpreters'. The constellation general model template implemented in CONSIM™ is based on a centralised shared memory approach.

The *Core Simulator* is the software module devoted to the control of the actual simulation process and its functional components are shown in Figure 5.9. Only the O&M simulation module has been developed, as yet.

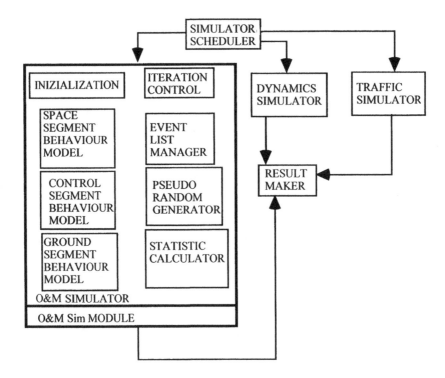

Figure 5.9   CONSIM™ model and core simulator architecture

A discrete events simulation mechanism has been adopted for the O&M simulation engine, currently being implemented, whilst an early design phase is ongoing for the selection of the best model of computation (i.e. discrete events, data flow, analytic models) to be used for the orbital dynamics emulator. Since several commercial and custom software are available for traffic simulation, the baseline approach in CONSIM™ has been to consider the integration with an 'off-the-shelf' existing tool. As highlighted in Figure 5.9, the overall O&M model of the constellation has been partitioned into three independent sub-models representing the behaviour of the space segment, control segment and ground segment, respectively.

5.4.1 A Communicating Process Approach

O&M aspects are efficiently modelled as a classical reactive systems because they represent the system reaction to sudden events. As a consequence, the selection of a Communicating Processes Model of Computation (MoC) appears to be the most appropriate.

In the CONSIM™ implementation of this MoC, independent actors execute non-interruptible tasks activated by the reception of incoming messages. More precisely each actor consists of a 'one-state', multi-thread finite state machine responsible for updating the state variables of the model according to the occurring event. For example, Figure 5.10 illustrates the processes (actors) and the messages used in the space segment sub-model of the O&M simulator.

The communicating processes are executed according to a discrete-event simulation tech-

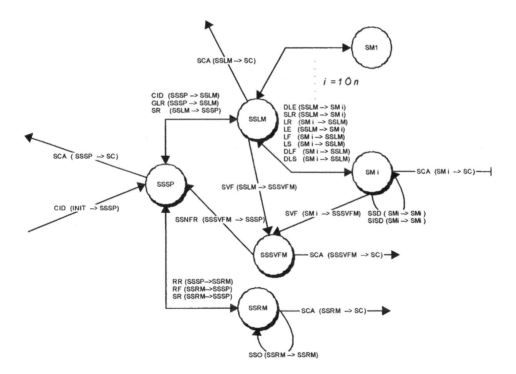

Figure 5.10   Processes of the space segment O&M model

nique – meaning that each message is time-stamped and all the message are time-ordered in a event list waiting for execution. The first message in the list is, by definition, the next occurring event and its execution fires the relevant thread in the destination actor. The management of the event list must solve the problem of possible synchronous events by either a proper priority assignment policy or avoiding 'zero-delay' message loops. It is worth noting that, by considering the actors as 'black boxes' with proper interfaces for interchanging messages, different paradigms could be adopted (e.g. data flows) by any actor.

5.4.2 Customisation of the Model and the Control Scheme

The customisation of the general purpose model template simply requires the user to provide a textual description (ASCII files) of the system behaviour by choosing among a variety of tasks and sub-procedures provided as a standard library. This is performed by using a sub-set of the ITU-T SDL standard language for the definition of the tasks and procedures executed by any actor.

The *basic user* is allowed to choose among the standard library tasks. The C++ code actually executed by the simulator engine is dynamically linked after completion of the interpretation phase; the *expert user* is also allowed to extend the standard library by creating his specific algorithms and editing the relevant C++ code. This clearly suggests a rebuilding of the environment by means of recompilation and linking phases.

Since the collection of statistical data is considered by the environment as a process, the textual description approach can be adopted also for the specification of the rules driving the simulation. This means that any statistical observation is fired by the reception of a specific message, activating procedures responsible for the monitoring of the model status variable.

It is important to highlight that in general, the formal description of communicating processes does not necessarily define any specific MoC, but constitutes a powerful tool to handle the system model complexity (i.e. partitioning). In this view, the communicating processes, intended as a set of independent actors exchanging information by means of messages over logical channels, are a suitable framework for distributed and parallel simulation. With reference to the core simulator architecture shown in Figure 5.9, the transition towards a parallel simulation scheme would require the removal of the centralised 'simulator scheduler' and the assignment of the message handling function to any independent 'sim module' interface.

The selection of the pseudo random generation algorithm to be used in CONSIM™ was carried out after a survey on the mathematics undergoing the simulation of the uniform distribution of random numbers and the analysis of the most studied and used algorithms. This lead to the choice of the Park and Miller algorithm, calculated according to the Schrage method.

5.4.3 Development Status

The first phase of the CONSIM™ framework design and development has been completed with the integration of the basic and most critical elements of the Core Simulator. In particular, the following modules have already been implemented and tested:

- The discrete-event scheduler inclusive of the message priority solver (event list manager);
- The pseudo-random number generator module with a selectable generation algorithm (Park and Miller as a default);
- The Reference Model builder (able to instate all the needed object-oriented model data derived from the architecture of the system to be simulated);
- The Reference Model I/O standard interface (to access and update the status variables of the model);
- The SDL interpreter and the related model behaviour builder (to interpret the SDL files describing the model behaviour and to link the library procedures to be executed during simulation);
- The Reference Model Template for the O&M aspects of the space segment (including architecture and communication processes behaviour);
- A subset of the O&M Core Simulator libraries;
- The Statistical Observer process;
- The System Architecture Interpreter;
- The preliminary GUI for Unix and Windows environments for testing purposes.

5.5 AristoteLEO

AristoteLEO is a simulation tool for satellite systems, developed for the performance evaluation of different routing algorithms in several constellations. The evaluation is performed in terms of blocking probabilities, throughput, delay (propagation and queuing), and system utilisation.

The basic need for developing AristoteLEO was to fully control the simulation tool. This demand originates from the fact that is impossible (in terms of simulation speed) to simulate satellite systems that serve large volumes of traffic, by using any of the simulation tool available on the market. AristoteLEO managed to carry out simulations in relatively small time periods.

5.5.1 The User Interface

AristoteLEO was not designed for distribution but for academic purposes; thus it comprises a simple user interface. This interface consists of two components, one related to the input/output files, and the second one related to the run-time input of parameters.

5.5.1.1 Input Parameters

Parameters are imported from input files and at run-time. There are three simple ASCII input files. The first one is devoted to the description of the constellation, the second one to the connectivity of the space segment, and the third one to the geographic traffic models that have to be examined. At run-time the user chooses the total simulation time, the implemented routing algorithm (among two different, at the current time) and the cost function used for each algorithm (among two cost functions for each algorithm).

5.5.1.2 Output Parameters

The output parameters are in ASCII file format. Several files are generated by the simulation, containing statistics about:

- Earth traffic sources
- Satellites
- Intersatellite links
- Up-down links
- The whole network

These statistics include blocking probabilities, forced blocking probabilities, successful handoff/handover probabilities, number of serviced calls, loading of ISL's and UDL's and propagation and queuing delays.

5.5.2 The Architecture

The AristoteLEO tool was realised by using the MODSIM language. MODSIM is an object oriented programming language specialised for event-driven simulations, unlike general purpose object oriented programming languages. The simulation engine of MODSIM is fixed, and it is implemented by scheduling events of each object separately. In this way MODSIM simulation engine overcomes the computational power needed for sorting simulation events, especially when events of different time scales are sorted together. The disadvantage of this simulation engine is that it becomes time consuming when the simulation involves a great number of objects that produce events.

The performance of MODSIM in terms of other common operations can be considered comparable with the performance of Visual C++. A special attention must be made in the manipulation of files. Although MODSIM provides a very sophisticated way for creating and manipulating files, the related operations are time consuming.

5.5.2.1 The Operating System Where it Runs

AristoteLEO aims to be portable in systems operating in any version of Windows and Windows NT. A small MODSIM dynamic link library (400 kbytes) moved together with AristoteLEO is only required. The input and output files related with the operation of AristoteLEO are in ASCII format and therefore portable. Figures 5.11–5.14 show the most relevant block diagrams.

5.5.2.2 The Performance

The performance of AristoteLEO strongly depends on the amount of traffic that is simulated and the specification of the simulated constellation. For the simulation of the Iridium system, at call level, with input traffic of 30,000 Erlangs, AristoteLEO performed with a correspondence of 1–6 for simulation and real time, on a Pentium III at 450 MHz machine.

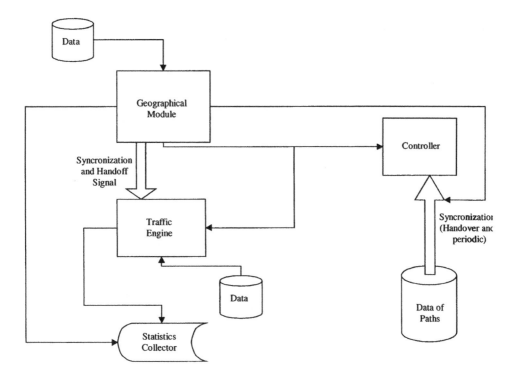

Figure 5.11   AristoteLEO

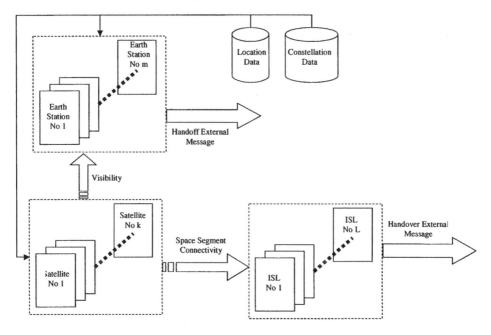

Figure 5.12   Geographical module

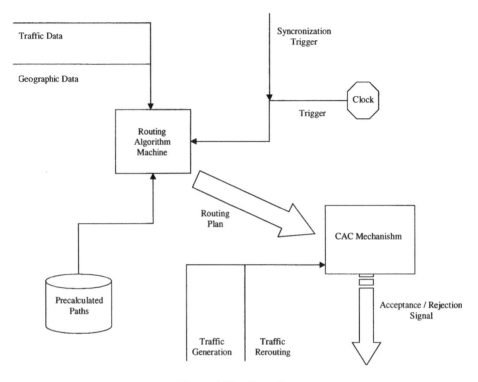

Figure 5.13   Controller

5.6 SEESAWS: An Ambitious Concept

Based on the specific experience achieved with several custom developments in the fields of the simulation and performance evaluation, an ambitious idea has been developed, (originally among some of the companies and institutes contributing to the Action COST 253 and then shared with other external companies). The objective is to provide the scientific community with a common simulation framework tailored to requirements of simulating large scale satellite networks based on constellation. SEESAWS (Simulation and Emulation Environment of SAtellite constellation for Wide-band Services) is just a proposal for a challenging project aiming to design, develop and validate a modular simulation and emulation environment to be made available to the entire satellite community for downloading free of charge as an open source software. SEESAWS could constitute an effective modelling framework for a large User Group, initially constituted by members of the SEESAWS team and then enlarged to the whole Satellite Community, in order to exchange models, compare results, contribute in the development of model libraries.

The SEESAWS framework would allow to evaluate the performance of satellite networks by providing a valuable insight into the achievable performance of future satellite systems, in terms of quality of service, mainly from a network perspective standpoint. By using the simulation/emulation environment, system designers would be able to verify how different design solutions would impact on the system performance so minimising the risks from the

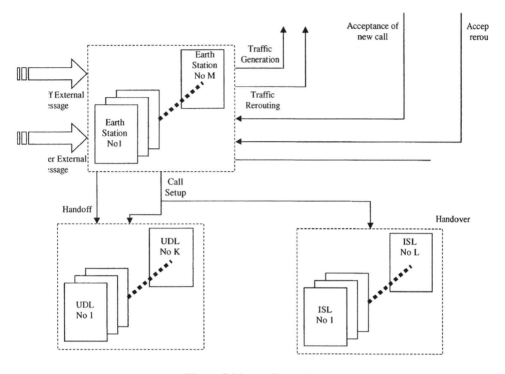

Figure 5.14   Traffic engine

early stage of the design phases. On the other hand, potential satellite service providers and investors could perform comparative analyses among different candidate system solutions in order to better understand if their expectation can actually be met.

Several simulation tools have been developed to assist terrestrial network design, planning and performance analysis, however, there is no specific tool available for the analysis of satellite networks and their particular characteristics.

The main objectives of the SEESAWS concept are to define, implement and validate a modular and integrated modelling and simulation environment to be used by both satellite network manufacturers and satellite network operators. Manufacturers could use this flexible framework to compare different system design alternatives and to obtain some useful feedback from the early phases of the design. Operators could use the framework to perform comparative performance analysis on the considered system in order to better balance their costs and benefits and to plan accordingly. Researchers could benefit from the presence of a common simulation framework to exchange models and libraries and to perform comparative performance analyses. The resulting environment should provide the capability of modelling and simulating:

- Different satellite network topologies based on an user provided system description;
- Aggregated traffic offered to the satellite network by any specific geographical region on the Earth based on traffic and mobility models, by taking into account predicted market shares and using a Satellite-Topographical and Geographic Information System (S-TGIS);

- Basic components of a satellite, such as the access subsystems, the inter-satellite link subsystems, on-board routing and processing subsystems, from a network standpoint;
- Time-variant topologies of the satellite network and the relevant implications (i.e. hand-off, link availability, congestion, fading, etc.);
- Congestion in the global network, where 'global' includes both the ground level and the constellation level;
- Reliability of the satellites.

The reference architecture of the SEESAWS framework is shown in Figure 5.15 and consists of the following five main components (having a clear correspondence with the physical structure of a typical satellite constellation network) suitable for parallel and modular design and development:

- Traffic Simulator;

    Regional Traffic Simulator;
    Traffic Statistics and Control;

- Satellite Simulator;

    Satellite On-Board Router Simulator;
    Satellite Ground Access Simulator;
    Satellite Inter Satellite Link Simulator;

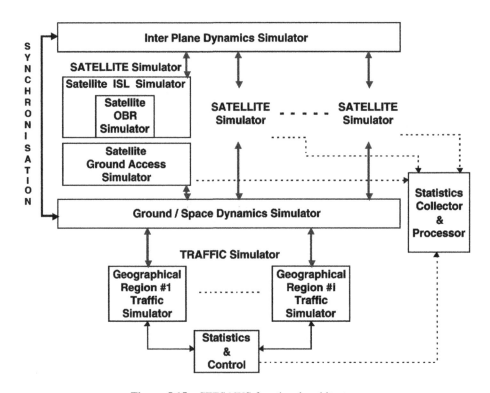

Figure 5.15 SEESAWS functional architecture

- Ground / Space Dynamics Simulator;
- Inter Plane Dynamics Simulator;
- Statistics Collector and Processor;

In the following sections a short description of these functional components is provided.

5.6.1 Traffic Simulator

This component deals with the simulation of the aggregated traffic offered, by any specific geographical region on ground, to the space segment, it also manages the received traffic and performs all the relevant statistical evaluations (Statistics and Control module).

Establishing a global traffic model is a difficult task. The first challenge is to develop a model able to cope with the huge number of consumers. With that respect, it is probably necessary to develop a 'two level' model where the studied area is modelled exactly, while the remaining areas are approximated. The second challenge is related to the range of services to be provided by the simulated network: a consistent global model must be crafted by taking into account how the different traffic characteristics may vary for different services.

From a simulation technique standpoint, the traffic simulator is one of the corner stones since it has an important influence on the global performance of the entire simulation environment. Indeed, traffic simulation, in order to be as accurate as possible, must be performed at the packet level. Unfortunately, considering broadband services as well as the network wide coverage, packet level simulation is not tractable; as a consequence, in order to decrease the simulation load and therefore shorten the simulation time techniques, such as packet aggregation or traffic behaviour simulation have to be adopted.

The traffic simulator component is intended as a sort of modular library of different user-defined traffic generators (TG), created for the generation of both real-time and non-real time data. Each traffic generator has its unique set of characteristics, such as: type of traffic generated (voice call connection, Poisson, fixed rate, fractal, 2-state Markov modulated Poisson, etc.); data destination address; data generation time distribution; data generation time jitters; packet length distribution; and contents pattern, etc. By properly combining the various types of traffic simultaneously generated by the TGs, it is possible to simulate arbitrary complex traffic patterns.

5.6.2 Satellite Simulator

This component consists of three modules: the satellite Ground Access (GA) simulator, the satellite On-Board Router (OBR) simulator, and the satellite Inter-Satellite Link (ISL) simulator.

5.6.2.1 Satellite On-Board Router Simulator

The OBR simulator is the component implementing the following functions:

- Switching;
- Congestion control;
- Call admission control;
- Down-beam routing;

- Downlink access technique;
- Uplink access technique.

The switching function is in charge of forwarding the data either to the next satellite (in case of a multi-hop connection) or to a ground station, according to the destination of the data. The switching function relies on the routing tables in order to take the right decision. These tables are computed by the ISL routing algorithm.

Monitoring the status of the buffers and queues in the satellite is the role of the *congestion control* function. If congestion occurs, congestion control techniques must be activated, such as to discard packets and to notify the traffic sources in order to throttle down the packet emission rate.

In order to reduce the possibility of congestion, the *call admission control* acts as preventive measure evaluating whether or not incoming traffic can be handled by the satellite. *Down-beam routing* is the process by which the satellite selects the beam supporting the channel between the destination satellite and the destination ground station. Down-beam routing is varying over time due to the movement of the satellite. When cell frequency pattern re-allocation occurs down-beam routing may also be invoked in order to re-compute the mapping between the channels and the beams.

The *up/downlink access techniques* implement the access scheme (TDMA, FDMA) which is used in the satellite for the up- and downlinks. The uplink access technique is the receiving entity of the uplink access technique implemented in the ground. The downlink access technique is the transmitting entity communicating with the receiving entity located in the ground station.

5.6.2.2 Satellite Ground Access Simulator

It is the component devoted to the modelling and simulation of the traffic streams received by any satellite uplink spot and transmitted by any satellite downlink spot. The functions are the following:

- Uplink routing,
- Down link routing,
- Congestion control,
- Uplink access technique,
- Downlink access technique.

The UDL routing function is performed in this module. The Uplink (UL) routing is the process by which the source ground station selects the source satellite used to forward the packets of the connection, while the Downlink (DL) routing is the process by which the destination ground station selects the destination satellite from which the packets of the connection will arrive. Possible criteria used for the UDL routing may be: the resources availability in the satellite and in the ground stations; the minimisation of the hand-over rate on the UDL; and the quality of the communication between the ground station and the satellite.

A congestion control module is also included in this component. The congestion control handles data congestion at the ground station level. Congestion may happen due to the status of the resources in the ground station or in the satellite, as a result of a huge amount of traffic which cannot be let through. This module allows the experimentation of different congestion control techniques. If congestion occurs, appropriate actions are taken such as throttling down the sources. For example, the call block probability relevant to a call connection is an indicator of the congestion of the resources in the satellite currently illuminating a certain area.

The uplink and downlink access techniques are respectively the transmitting and receiving entities of the access techniques previously described. The satellite access techniques (TDMA, FDMA) should take into account fading problems due to rain, multi-path, and shadowing.

5.6.2.3 Satellite Inter Satellite Link Simulator

It is the component dedicated to the modelling and simulation of the routing policy implemented in the satellite. Once decided by the satellite OBR simulator that data must be addressed to another satellite, this module computes the optimal path (or at least one path) between two satellites. Possible criteria used are: the resource availability in the satellites, the minimisation of the hand-over rate; the quality of the communication among the satellites; and the length of the path. As a result, the optimal path is the one which fulfils the user requirements most accurately, whilst minimising the amount of used resources.

The main assumption is that, in most of the future systems the satellite is like an actual node of the space network. Therefore, it must be able to take independent decisions about the routing of the received packet according to the real time status of the network.

Routing is an important aspect of satellite constellations since the dynamic environment of the constellation calls for revisiting the routing architecture. Among others, one of the important issues is the hand-over management. Furthermore, different routing approaches currently exist in terrestrial networks. The most appropriate one must be chosen in order to fit with the environment featured in the constellations.

5.6.2.4 Ground/Space Dynamics Simulator

This component is based on the characteristic orbital parameters of the specific constellation considered. It simulates the effect of the motion of the satellite footprint by controlling the assignment of the traffic sources and sinks (associated to the ground regions) to the proper Satellite Ground Access Simulator.

5.6.2.5 Inter-Plane Dynamics Simulator

This simulator is also based on the characteristic orbital parameters, and simulates the discrete status variations in the link between satellites. Two different types of orbits are considered: circular orbit (which includes equatorial orbit, polar orbit, and inclined orbits), and elliptical orbit. A geostationary satellite is a special case of synchronous orbit – it is a satellite in circular orbit over the equator.

5.6.2.6 Statistics Collector and Processor

This module is not related to a specific component of the system but collects statistics from any of the other components. As an example, a list of some of the possible statistics the user could collect, is listed below:

- End-to-end delay, jitters, losses (QoS);
- Amount of data generated by a station, corresponding to specified types of traffic;
- Amount of data transmitted over a certain satellite, averaged over the constellation, averaged in a time interval, etc.;
- Amount of data lost due to ground congestion;
- Amount of data lost due to satellite congestion;
- Percentage of availability of the system according to different fading conditions, system faults, link availability, etc....;
- Call connection drop probability;
- Call connection block probability.

References

[1] OPNET Tutorial Manuals. MIL3, Inc.
[2] K. Sam Shanmugan, *BONeS Designer, Introductory overview*, COMDISCO Systems, Inc.
[3] C.H. Sauer and E.A. MacNair, 'Queuing network software for systems modelling', *Software-Practice and Experience* Vol. 9, No. 5, 1978.
[4] C.H. Sauer, E.A. MacNair and J.F. Kurose, 'The research queuing package: past, present and future', *Proceeding of the National Computer Conference AFEPS*, Arlington, VA, USA, 1982.
[5] A. Cohen and R. Mrabet, 'An environment for modelling and simulating communication systems: application to a system based on a satellite backbone', *International Journal on Satellite Communications*, Vol. 13, pp. 147–157, 1995.
[6] C.D. Pham, J. Essmeyer and S. Fdida, 'Simulation of a Routing Algorithm using Distributed Simulation Techniques' *Euro-Par'97 Conference*, Passau, Germany, 1997.
[7] M. Annoni – CSELT, 'Reliability modelling for constellation systems', COST253 Working Document TD(98) XX.
[8] P.A. Muller, 'Instant UML', Wrox Press.
[9] M. Campione, K. Walrath and A. Huml, *The Java Tutorial*, 2nd Edition, Reading, MA, Addison Wesley, 1998.
[10] M. Campione, K. Walrath and A. Huml, *The Java Tutorial continued*, Reading, MA, Addison Wesley, 1998.
[11] D. Lea, *Concurrent Programming in Java*, Reading, MA, Addison Wesley.
[12] R.C. Lee and W.M. Tepfenhart, 'UML and C++', Eaglewood Cliffs, NJ, Prentice Hall, 1997.
[13] E. Ferro and L. Franck, 'TD(98)019a – Simulation of the satellite constellation and the ground stations that access it', CNUCE/CNR, ULB Bruxelles University.
[14] E. Ferro and L. Franck, 'TD(98)001a – GaliLEO: a simulator of the satellite constellation and the ground stations that access it', CNUCE/CNR, ENST.
[15] R.M. Fujimoto, *Parallel and Distributed Simulation Systems*, New York, Wiley.
[16] C.L.R. Guthrie, *Java2 Performance and Idioms Guide*, Eaglewood Cliffs, NJ, Prentice Hall.
[17] G. Maral, J.J. De Ridder, B.G. Evans and M. Richharia, 'Low Earth orbit satellite systems for communications', *International Journal of Satellite Communications*, Vol. 9, pp. 209–225, 1991.
[18] J. Misra, 'Distributed discrete-event simulation', *ACM Computing Surveys*, Vol. 18, No. 1, 1986.
[19] S.K. Park and K.W. Miller, 'Random number generators: good ones are hard to find', *Communications of the ACM*, Vol. 31, No. 10, pp. 1192–1201, 1988.
[20] M. Annoni and S. Bizzarri, 'A simulation environment for the evaluation of maintainability strategies in telecommunication networks based on satellite constellation', *Mobile and Personal Satellite Communications 2; Proceedings of EMPS '96*, pp. 233–246, 1996.
[21] M. Annoni, S. Bizzarri and F. Faggi, 'Performance evaluation of satellite constellations. The CONSIM simulator concept and architecture'. *Third European Workshop on Mobile/Personal Satcoms, EMPS '98*, Venezia Lido, Italy, 1998.

[22] M. Annoni, S. Bizzarri and F. Faggi, 'CONSIM: a flexible approach to satellite constellation simulation', *DSP'98, 6th International Workshop on Digital Signal Processing Techniques for Space Application*, 1998.

[23] M. Annoni and S. Bizzarri, 'An integrated modelling methodology for the simulation of the life cycle of satellite constellations', *Joint COST 252/253/255 Workshop*, Toulouse, May 1999.

[24] M. Annoni, 'TD(98)018–CONSIM: concept and architecture' CSELT.

[25] NS web reference: www.isi.edu/nsnam

[26] K. Fall and K. Varadhan, *NS manual/NS notes.*

6

TCP/IP Over Satellite

Since the 1960s, the use of satellites has grown rapidly, with applications ranging from telephony, TV broadcasting, and navigation to personal and mobile communications. The type of satellite used to provide such applications has also evolved from simple geostationary solutions to sophisticated satellites, utilising a variety of orbital constellations.

The Internet has also been around since the sixties, however, its mass market appeal has only really taken off in the last few years. The growth of the Internet has resulted in its protocol design being very much based on the needs of terrestrial networks, with little consideration for its potential use over satellite. Here, a comparison can be made with mobile communications, where the vast majority of effort has focussed on the development of terrestrial mobile networks, such as GSM (Global System for Mobile communications) and UMTS (Universal Mobile Telecommunications System).

The coverage characteristics of satellites make them especially attractive for deployment of services in hard-to-reach areas and places where it is uneconomical to build or roll-out terrestrial networks. Hence, it can be seen that a satellite network can either be used as an access link to terrestrial networks or for interconnecting links separated by large distance [12]. Besides that, satellite systems can also be used to reduce network congestion in terrestrial systems by providing a means to off-load traffic.

As world-wide communication becomes more and more important, satellites are increasingly being used to deliver Internet traffic, offering high speed Internet access and multimedia applications to complement the services provided by the terrestrial networks. A number of recent studies have clearly shown that Internet load is primarily determined by Transmission Control Protocol (TCP) connections resulting from the World Wide Web (WWW). For example, in [4] it is shown that TCP traffic in a large University campus access link amounts to 88.78% of bytes transmitted, 72.91% of which (91.71% of the total number of TCP connections) are as a result of WWW connections. Moreover, in [5] it is observed that the percentage of TCP traffic equals 99% of bytes transmitted, 82.8% of which (96.9% of the total number of TCP connections) are due to WWW connections. Such a high percentage is also reported in a number of measurements performed in a wide variety of academic and industrial settings [8,29].

In the following, a discussion on the present state-of-the-art, in terms of TCP over satellite, is presented. Much of this work is based on the RFC (Request for Comment) documents provided by the Internet Engineering Task Force (IETF), the body responsible for the development of the Internet. Initially, the mechanisms involved in establishing and maintaining a connection are described. This is followed by a review of the particular characteristics pertinent to satellite communications that require special consideration, in comparison to conventional terrestrial network implementation. An analysis by simulation is then presented, the aim of which is to investigate the performance of TCP over different satellite constellations under various operational scenarios. Finally, in recognition of the convergence between fixed and mobile networks, an overview of the current status of Mobile IP, with particular emphasis on satellite aspects, is then discussed, followed by concluding remarks to complete the Chapter.

6.1 Transmission Control Protocol

6.1.1 Introduction

TCP, being a part of the Internet protocol suite, provides a reliable delivery of data segments. Most Internet applications, such as e-mail, file transfer protocol (FTP), Web browsing (HTTP), and remote connectivity services (X-Window), use TCP to move data between clients and servers.

Reliability between TCP users is ensured by using acknowledgements for every data segment sent to the destination. Any missing segments, indicated by the reception of a duplicate acknowledgement (DACK), are retransmitted.

Being a connection-oriented protocol, TCP achieves a certain level of reliability that the Internet Protocol (IP) does not provide. TCP itself is not present within the network level, but operates at the end of the connection, at transport level, with IP providing the interface to the network.

Before any data can be transferred between any client and server using TCP, a connection must be established using a three-way hand shake. Once this is achieved, data can be sent across the network.

When establishing a new connection, TCP will be unaware of the current network situation, in terms of traffic load. In order to prevent an inappropriate amount of traffic being injected into the network, four standard TCP congestion control algorithms are used. These are *Slow Start*, *Congestion Avoidance*, *Fast Retransmission and Fast Recovery*. These algorithms ensure that all segments arrive in order, and any lost or corrupted data are re-transmitted. TCP also uses two variables, Congestion Window (*CWND*) and Slow Start Threshold (*ssthresh*), to accomplish congestion control.

6.1.2 Slow Start and Congestion Avoidance

Without TCP mechanisms, a client could inject any number of segments into the network. This could prove disastrous for a congested traffic condition. A TCP transaction begins with a connection establishment phase, followed by slow start. Slow start is used as an initial start up process for any TCP connection. It is a precautionary mechanism which prevents a user from overloading the network.

Once a connection is established, slow start initialises the client's CWND to one segment and *ssthresh* to the receiver's advertised window [28]. The slow start algorithm makes transmission from the server be "clocked" by ACKs from the client. Hence, it can be seen that slow start makes TCP transmission behave in a stop-and-go manner. For each ACK received during slow start, two TCP segments can be sent, until the window size reaches a

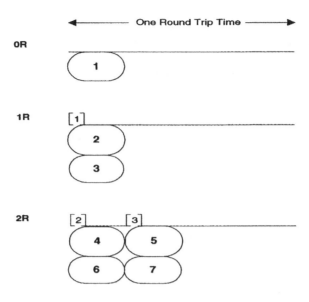

Figure 6.1   Slow start process

value (*ssthresh*) that allows for steady-state transmission. Such a value depends on the buffering in the bottleneck link and the socket buffer size of the TCP connection. When steady-state occurs, the congestion avoidance phase takes over, assuming that no loss of TCP segment occurs. Figure 6.1 shows the whole process of slow start.

Congestion avoidance increases the CWND size at a much slower rate than the slow start phase. For each ACK received, CWND is increased by 1/CWND.

Congestion control algorithms can cause poor utilisation of the available channel bandwidth when used on a long delay path. For an *ssthresh* window size of *W* segments and a round trip time (RTT) of *R*, as defined by the satellite or terrestrial network, the amount of time required for TCP to reach the *ssthresh* window size level is given by the following equation:

$$\text{Start Up Time} = R \log_2 W \tag{6.1}$$

The longer the RTT, the longer it takes TCP to get itself up to speed. This also means a long period of idleness when no other users are in the network.

6.1.3 Fast Retransmission and Fast Recovery

It is possible during transmission that one or more TCP segments might not reach the destination. Packet losses are detected either by a time out or the reception of a duplicate acknowledgement. When TCP uses time out mechanisms, this results in *CWND* size being reduced to one TCP segment size and *ssthresh* being reduced to half of its current size. The slow start algorithm then follows, i.e. the transmitter is reset to the initial connection state.

When the loss of TCP segments is not due to network congestion, in order to avoid the unnecessary process of going back to the slow start procedure each time a segment is lost, TCP Reno introduces the process of Fast Retransmission [21].

The Fast Retransmit algorithm uses duplicate ACKs to detect the loss of segments. Once three DACKs are received at the sender, TCP immediately retransmits the missing segment without waiting for the time out to occur, so that the receiver recovers promptly.

Once Fast Transmission is used to retransmit the missing data segment, TCP can use its Fast Recovery mechanism, which will resume the normal transmission process via the congestion avoidance phase, instead of slow start as before. However, the size of both *CWND* and *ssthresh* will be reduced to half their previous values after normal operation resumes.

Recall that both Fast Retransmission and Fast Recovery will start only when three dupacks are received within the normal time out period, indicating that the lost segment was not due to network congestion. However, if segment loss is detected through a time out, slow start is initiated.

Fast Retransmission works well for the loss of a single segment per RTT, however, this is not the case if more than one segment is lost during one RTT. In such a situation, Fast Retransmission would only detect the first lost segment, while the remaining will be detected via time out mechanisms and therefore bring the whole TCP process back to the slow start phase, as before.

Finally, note that the loss of an ACK segment is not as harmful as the loss of a data packet, since ACKs are cumulative. However, if several data packets are transmitted in a single burst then the connection is likely to go back to the slow start phase.

To improve TCP performance for this situation, Selective Acknowledgement (SACK) is proposed [18].

6.1.4 Selective Acknowledgement (SACK)

TCP, even with Fast Retransmission and Fast Recovery, still performs poorly when multiple segment losses occur within one RTT. This is due to the fact that TCP can only learn of a missing segment per RTT, due to the lack of cumulative acknowledgements. This limitation drastically reduces TCP throughput.

Selective acknowledgement mechanisms let the server know exactly which packets have been lost, rather than providing a cumulative ACK. Since the error recovery loop encompasses a satellite round trip, it is intuitively appealing to make the server retransmit all the lost packets as soon as it receives a selective confirmation of correctly received packets, instead of waiting for several cumulative ACK packets to progressively fill the gaps in the receiver window [2,19].

In order to use SACK, a negotiation is performed during a connection establishment using a two-byte option called SACK-permitted together with the SYN segment. A SACK option format is shown in Figure 6.2.

The above SACK option format allows any missing data segment to be identified using the block field. Each block represents data segments that have arrived safely. This is identified by the left and right edges of the block. Any segment that is not in the covered range will need to be retransmitted.

The number of blocks that this option can carry is limited by the 40 bytes available in the TCP options field. For a total of 'n' blocks, the total length is $(8 \times n + 2)$ bytes. The additional

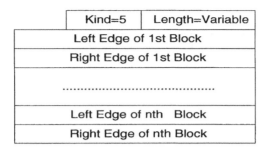

Kind=5	Length=Variable
Left Edge of 1st Block	
Right Edge of 1st Block	
....	
Left Edge of nth Block	
Right Edge of nth Block	

Figure 6.2   TCP SACK option format

two bytes come from the kind and length fields, which each take up one byte. This limits the maximum number of blocks to four. However, as SACK is likely to be used with the time stamp option, which would take an additional ten bytes, a maximum number of three blocks is allowed.

6.2 The Effects of Satellite Networks on TCP Performance

6.2.1 Overview

For several years, significant effort has been directed towards making TCP perform more efficiently over satellite networks. TCP, designed primarily for terrestrial networks, does not perform well over satellite. Some of a satellite network's characteristics that degrade the performance of TCP are long delay path, large delay-bandwidth product, asymmetric channels, channel quality (BER), variable round trip time, intermittent connectivity and limited bandwidth. These limitations are considered further in the following sections.

6.2.2 Large Bandwidth Delay Product

The bandwidth delay product (BDP) defines the maximum amount of unacknowledged data that a client can have in flight or can inject into the network. Such a network is called a *long fat network* and can be specified by the following equation.

$$BDP = Bandwidth \times Round\ Trip\ Time \tag{6.2}$$

It is essential that the TCP window size must reach the size of the bandwidth delay product in order to fully utilise the channel. TCP is unable to achieve this due to the limited window size of 65536 bytes. This limitation is due to the 16 bits available in its window size field. (Most Internet servers have a maximum advertised window size in the order of 32 kbytes [4], due to memory limitations in the presence of multiple concurrent TCP connections for the most popular servers.)

Typical values for TCP utilisation are shown in Table 6.1.

It can be seen that the percentage utilisation drops dramatically with a large increase in channel capacity.

In the satellite case, the bandwidth delay product is extremely large for geostationary

Table 6.1   TCP channel utilisation

Network	Bandwidth (Mbits/s)	RTT (ms)	Bandwidth Delay Product (bytes)	TCP Utilisation %
T1	1.544	500	96,500	68
T3	45	500	2,812,500	2.3

(GEO) satellites, due to the round trip time. As for Low Earth Orbit (LEO) satellites, the round trip time is reduced considerably but still the bandwidth delay product is significant.

Figure 6.3 shows the data transfer phase dynamics between server and client in a high bandwidth delay product scenario, where both client and server are connected through a satellite link via the corresponding satellite gateways. Note that the TCP model presented in Figure 6.3 corresponds to the cyclic model of 2^n packets per cycle window(n) that is commonly adopted in order to characterise TCP behaviour in large bandwidth delay product scenarios.

To fully utilise the channel capacity of a long fat network, the use of TCP extension options is required. These options include windows scaling, time stamp, and Protection Against Wrapped Sequence Numbers (PAWS).

6.2.3 Long Delay Path

In order to ensure that all data segments sent by one end of the connection are received by the other, TCP uses retransmission time out (RTO) to indicate segment losses. This algorithm affects the TCP performance over satellite due to the long delay paths. According to equation (1), it can be seen that the time taken for TCP to reach the congestion avoidance phase directly depends upon the RTT, which in turn is used to calculate the RTO.

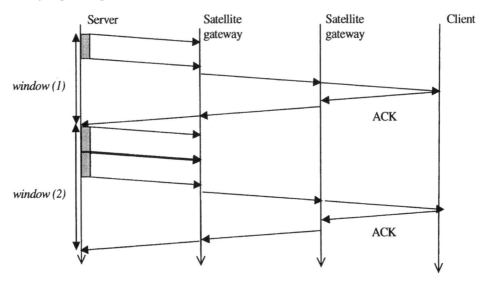

Figure 6.3   TCP connection data transfer phase

6.2.4 Bit Error Rate versus Congestion

Communication using satellites usually has a higher bit error rate (BER) in comparison to terrestrial links. This high BER arises from the fact that the signal has to travel over long distances and with limited transmitting power. This makes the signal susceptible to noise, which in turn causes error.

TCP assumes that all the dropped segments are the result of network congestion, and reduces its window size accordingly, before returning to the slow start phase. In doing so, the instantaneous data rate from the server is reduced, so that the load on the bottleneck link is alleviated and buffer overflow is avoided. However, if losses are not due to buffer overflow (congestion) but rather as a result of the physical channel conditions, then reducing the transmission window is of no use, since the channel BER will remain the same, irrespective of the instantaneous data rate. This is undesirable for satellite networks with long delays, as the time it takes to build up the window size to a reasonable level could result in long and unacceptable delays for some applications.

A way to solve such a problem, in conjunction with reducing high bit error rate, is to make each TCP to be more intelligent in distinguishing between a segment dropped due to error and those that are due to congestion.

Two approaches to deal with transmission error are proposed by [13]. Here, TCP can either be explicitly *told* or *inferred* that link errors are occurring. The *infer* method uses TCP ACKs as a means to provide the information, while *told* uses the data segment itself as the source of information for an occurring transmission error. However, there is a problem with this method, as the transmission layer discards the corrupted data packets before they are passed on to TCP. This prevents TCP from being able to extract any information from a corrupted segment.

As a means of counteracting the effects of the channel, a possible solution could be to employ link layer FEC techniques, which take care of the data packets which are lost due to bit error rate. Consequently, if a low BER is achieved, the TCP congestion control algorithm will only be triggered if the link is truly congested and not by misinterpretation of a packet loss event.

6.2.5 Link Asymmetry

TCP, being a sliding window protocol, relies on acknowledgements from the receiver to 'clock' data out from the transmitting source. TCP connections are highly asymmetric, with an inbound channel from the server to the client which is clearly dominant, since the outbound from the client is devoted to transmission of ACK and control messages only (for instance HTTP GET messages). Asymmetry severely affects throughput, and performance is limited by the lower capacity link. Figure 6.4 shows the probability density function and cumulative density function of the connection asymmetry, which is defined as percentage of bytes from client to server divided by total number of bytes exchanged in both directions.

Figure 6.4 shows that less than 25% of the connection in the traffic trace [4] produce more than 50% of the total bytes transferred in the client-server direction. The peak at 100% in the histogram shown in Figure 6.4 shows that connections are highly asymmetric in the client to server direction, for example, in FTP data transfers. It has to be noted that the histogram represents a normalised histogram such that a 100% asymmetry can be regarded as a 'close to 100% asymmetry'.

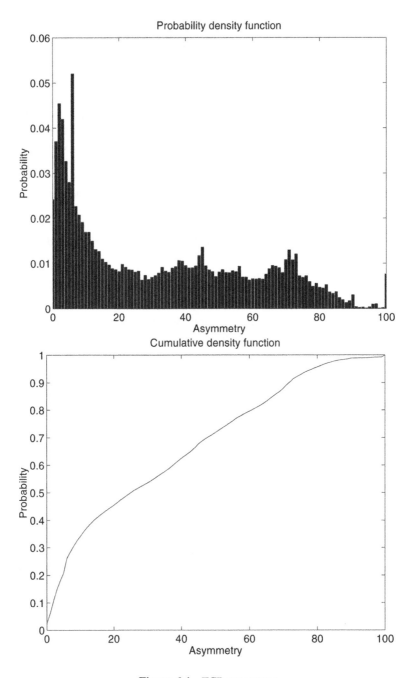

Figure 6.4   TCP asymmetry

In terms of capacity, communication channels between satellite and the ground are usually asymmetrical in character. It is usually the downlink (from satellite to ground) that has a higher capacity in a VSAT (Very Small Aperture Terminal) Network. If the server is attached to a Fixed Earth Station (FES) or gateway, then bandwidth will be highly utilised. Furthermore, data packets will not suffer from the packet loss conditions which are likely to occur in the outbound direction from a user station, which will be devoted to transmission of ACK packets. However, since the data rate from the server to the client is determined by the arrival rate of ACKs from the client, it can be seen that the client outbound channel may become the limiting factor for performance.

For example, a communication set up using a VSAT with a typical capacity of 64 kbit/s (8 kbytes/s), would limit the throughput of the downlink (T1) to approximately 690 kbit/s for an ACK of size 100 bytes made on an alternate data segment of size 536 bytes that arrives at the receiver. This limited reverse channel capacity can behave like a 'bottle-neck' if the reverse channel cannot cope with the rate of acknowledgement generated by the receiver. A possible solution is to use the delayed acknowledgement option, which could ease the amount of generated ACKs in the reverse link. However, this would mean a longer slow start start up time and results in poor utilisation of bandwidth.

6.2.6 Variable Round Trip Time

A variable round trip time is a major concern for TCP, when using a dynamic network topology such as a LEO satellite network. The RTT time in this case varies as the satellite moves across the sky. This is due to the increase or decrease in path length of the satellite and variation in network topology.

The Retransmission Time Out (RTO) calculation is based on the measured RTT [28]. As the RTT experienced by a TCP connection over a LEO satellite will vary significantly over time as the satellite moves, it is usual for the estimated RTO to be either underestimated or overestimated. If RTT varies such that the estimated RTO value is far less than the actual delay that a LEO satellite experiences, a premature RTO could occur, triggering an unnecessary slow start. This results in a reduction of throughput [9].

6.2.7 Adaptation of Access and Satellite Backbone Networks

When more than one user terminal (for example, a PC) is attached to a FES, the problem arises of adapting the various heterogeneous network segments along the path from the client to server. As stated previously, the bandwidth delay product of a satellite link is large in comparison to that of the terrestrial segment from the FES to the user terminal. The terrestrial connection can be realised by means of a narrowband ISDN link or analogue modem, for example. Since the TCP congestion control mechanism will adapt to the bottleneck link between source and destination, the connection throughput will be determined by the access link (e.g. ISDN) and the satellite bandwidth will be under utilised. Furthermore, if the access network is congestion prone, then a packet loss in the access network forces a retransmission from the server, which suffers from the round trip delay imposed by the satellite segment. Alternatively, the lost packet could be retransmitted from the FES, thus avoiding the performance penalty of a much higher RTT loop. In order to do so, split TCP connection mechanisms have been recently proposed, which perform segmentation of the TCP connection into

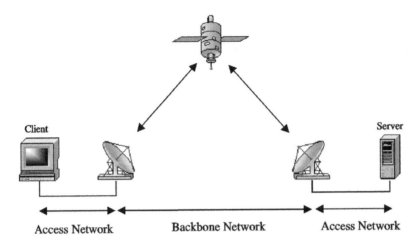

Figure 6.5   Split segment network architecture

two separate parts, access and backbone, whose transmission parameters are adjusted to the particular features of each network side. For example, a larger window size is required in the satellite segment and not in the access segment. In the server side, duplicate ACKs can be sent from the server FES and not from the client at the other end of the satellite segment, thus decreasing the chances of entering the slow start phase.

The split segment network architecture is shown in Figure 6.5, while Figure 6.6 provides a graphical illustration of the split TCP connection dynamics.

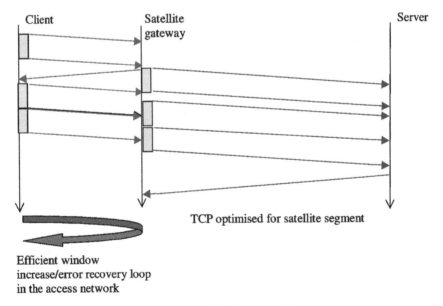

Figure 6.6   Split TCP connections

6.2.8 Intermittent Connectivity

Severe loss detection in TCP is performed by means of time outs, for which an estimate of the round trip time between client and server is required. Intermittent connectivity could be a serious problem when using TCP over satellite in the future, where large and complex satellite constellations are involved. Here, it would not be just complexity of the satellite constellation that would have to be dealt with but also the complex dynamic topologies involved. In the case of a LEO satellite network, both the satellite motion and the inter-satellite dynamic routing provide delay variations which are difficult to capture by the standard TCP RTT estimation procedures.

For a LEO satellite network, satellite visibility over a particular location is typically about 10 to 15 minutes. In order to maintain a TCP connection, frequent handover procedures are required between satellites, assuming no inter-satellite links [9]. These handover procedures are much longer than those experienced by a terrestrial cellular system.

TCP suffers from a reduction in throughput as a result of the loss of packets during a handover. As with the case of errors introduced by the physical channel, TCP assumes congestion in the network and reduces its congestion window and initiates slow start.

It is usual that during handover, a TCP connection could experience long periods of idleness. If this is for a long period of time, slow start is invoked. This is done as a precautionary measure, as TCP does not have up to date information about the present network condition. This would result in poor bandwidth utilisation, or in the worst case, a connection being aborted.

6.2.9 Bandwidth Limitations

Wireless communications are constrained by both power and bandwidth. The TCP protocol has a substantial bit overhead of 40 bytes per data segment. This can consume a sizeable share of the limited bandwidth available to the satellite. In order to solve this problem, a header compression is required or an increase in the segment size so that the overhead consumption is reduced.

6.3 Simulation Analysis

6.3.1 Network Architecture

In order to investigate the effect of TCP performance over satellite, a simple model of a client and server has been developed using the OPNET simulation software. The simulation set up is shown in Figure 6.7.

By connecting a client to a server via a representative Wide Area Network (WAN), incorporating LEO, Medium Earth Orbit (MEO), and GEO satellites, simulation can be performed to study the effect of running TCP traffic over the different constellations.

6.3.2 Simulation Results

6.3.2.1 Set Up
Using the simulation set up shown in Figure 6.7, the delay encountered in the point-to-point link between the client and server was varied to imitate the delays encountered in the

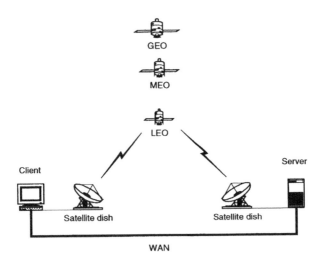

Figure 6.7 Simulation scenario

different satellite networks. For GEO and MEO satellites, a round-trip time delay of 480 ms and 138 ms were used, respectively. For LEO satellites, a transmission delay of 10 ms was applied. For comparison, a simulation was performed for a terrestrial network with a delay of 4 ms.

6.3.2.2 Effect of Delay

The performance of TCP over various delay networks is shown in Figure 6.8. By setting the same duration for transmission, the amount of data being transmitted via the various networks can be compared. Figure 6.8 clearly demonstrates that the performance of TCP suffers when operating in long delay networks. As the channel condition is not considered in this case, the BER is set to less than 10^{-12}. In the case where TCP segments are not dropped, one of the reasons for poor performance is the use of the slow start process at the beginning once a connection is set up. However, the major reason is still the long delay itself. It is clear from Figure 6.8 that standard TCP does not work well over a GEO satellite network.

However, little difference in performance can be detected between the LEO satellite and terrestrial networks. Many possible solutions have been proposed to improve the performance of TCP over GEO, as will be discussed shortly.

6.3.2.3 Effect of Channel BER Conditions

The performance of TCP over GEO satellite channels is achieved by adjusting the BER level. The results are shown in Figure 6.9.

TCP with both Fast Retransmission and Recovery and SACK performs better than standard TCP in the noisy channel. This is mainly due to their ability to recover from the loss of a segment without having to go back to the slow start process.

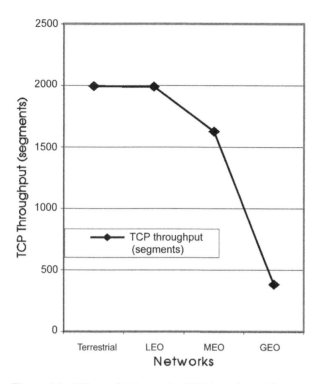

Figure 6.8   Effects of delay to the TCP throughput efficiency

6.3.2.4 Possible TCP Improvements over Satellite: Using larger Initial Window Size

A larger initial window size solution was proposed in [23]. Further evaluations on this proposal in [3] suggested an increase in the TCP initial window size from one segment to between two and four segments. The main reason being to reduce the amount of time spent during the slow start process. This is important in the case of long delay networks. The results are shown in Figure 6.10 indicating the level of improvement

In addition to increasing throughput, a larger initial window size also reduces the transmission time and improves the use of satellite bandwidth more efficiently by reducing the amount of time spent in slow start, as given in [3].

To demonstrate the reduction in the transmission time, a file size of 10 kbytes was transmitted with different initial window sizes. The variation was done from an initial size of 2–64 TCP segment sizes

Two segment window sizes were taken as the baseline for comparison, showing an improvement in the transmission time, as shown in Figure 6.10.

The simulation results also show a slight improvement as compared with the results obtained from [3], which uses a larger initial window size only for the start of the transfer, and after an idle connection uses a file size of 16 TCP segments for transmission, as specified in [23,3].

The improvement after the initial window size of 16 segments is due to the use of the larger link capacity of 64 kbps in the simulation, compared with a dial-up connection of 28.8 kbps

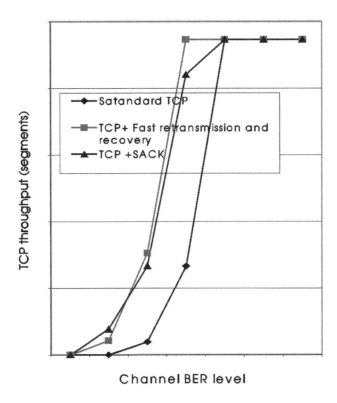

Figure 6.9 Effect of channel BER level

used by [3]. The transfer time in [3] increased as a result of segment loss, as the capacity of the dial up connection was being exceeded.

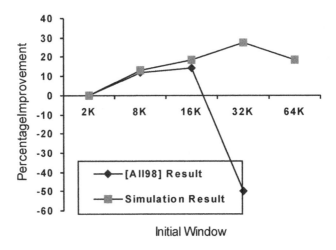

Figure 6.10 Transfer time improvements

6.3.2.5 Conclusion

It has been demonstrated that TCP performs poorly over long delay path and high bit error rate channels. This is because TCP assumes all data segment losses to be due to congestion, as stated previously. This reduces the ability of TCP to fully utilise the channel capacity and also increases the response time over a long delay path. The simulation results have shown as much as 40% drop in throughput when compared to a terrestrial network, and approaches zero as high bit errors are introduced into the link. In order to improve TCP performance, modification was done on TCP to use a larger initial window after a retransmission time out. The simulation results show an improvement, in terms of its throughput, by as much as 85% and an 18% improvement in its response time. It did not however perform well with multiple users. Further study is required to find an optimum window size that could be used without degrading TCP performance in a multiple user scenario or reducing TCP's effectiveness in preventing network congestion.

6.4 Fixed – Mobile Convergence

6.4.1 Introduction

Along with the Internet, the mobile communications sector can be considered to be the other major technological success of the last decade. Over the next few years, Internet and mobile technology will converge, such that services presently available at the desk will be also accessible while on the move. This forms the framework for the introduction of fourth-generation (4G) mobile technologies. Such a future service scenario requires several key technological areas to be developed over the next few years. So far in this Chapter, emphasis has been placed on fixed network performance at the TCP layer. The introduction of mobility into the future service scenario also requires consideration of the IP layer. In the following, aspects of Mobile IP, emphasising the requirements of mobility management and the relevance to satellite networks, are presented.

6.4.2 Mobile IP

6.4.2.1 Principles of Operation in Mobile IP Version 4 (MIPv4)

In normal IP operation, IP packets are routed to their final destination using an IP address, which typically carries information on a mobile node's (terminal) point of attachment. This point of attachment is identified by the network number and the host number contained in a mobile node's IP address. However, if a mobile node moves to a new location, which has a different network number, it becomes difficult for packets to be delivered to the node, as the network address of the new subnet will be different from the address of the subnet that the node was previously attached. To overcome this problem, the principles of Mobile IP [20] are introduced, where some new functional entities are added to the existing IP protocol, namely, a home agent and a foreign agent. A home agent is a router on the mobile node's home network that tunnels datagrams for delivery to the mobile node when it is away from home, whereas a foreign agent refers to a router on the mobile node's visited network that provides services to a registered mobile node.

Using Mobile IP, a mobile node is usually given a long-term IP address, known as the home address, when it registers with its home network. When this node moves away from home and is connected to a foreign network, it acquires a care-of-address (CoA) from the foreign network to reflect its current location, which it registers with its home network. Therefore, when the mobile node is away from home, it will maintain two IP addresses. When the node receives a packet, the home agent intercepts the packets/datagrams destined for the mobile node and then tunnels them to its CoA. Datagrams sent by the mobile node may be delivered to its destination without going through the home agent. This mechanism is illustrated Figure 6.11 and is known as the triangle routing method.

In MIPv4, two methods can be used to obtain a CoA. In the first method, the address is provided by a foreign agent. Therefore, the CoA of the mobile node is an IP address of the foreign agent. When the foreign agent receives a packet destined for the mobile node, it will decapsulate then deliver the packet to the mobile node. Using this method, the same CoA can be shared by all the mobile nodes attached to the foreign agent and therefore does not waste any IPv4 addresses. The second method involves acquiring the address through an external source such as Dynamic Host Configuration Protocol (DHCP) [22]. This type of address is also known as a co-located CoA. Using this address, packets can be tunnelled directly to the mobile node.

In MIPv4, an optional mechanism called route optimisation is introduced to eliminate the triangle routing scenario, in an effort to increase the efficiency of the Mobile IP protocols. When route optimisation is implemented, a corresponding node can deliver packets directly to the mobile node without any assistance from the home agent.

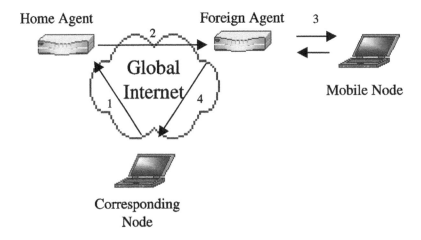

Figure 6.11 Mobile IP datagram flow

6.4.2.2 Changes with IPv6

The concept of Mobile IP in IPv6 (MIPv6) is very similar to MIPv4. However, due to the extended capabilities in IPv6 [25], the procedures for Mobile IP are slightly changed to accommodate IPv6. The specifications for MIPv6 are described in [11].

In MIPv6, the concept of the mobile node, home agent, home network and foreign network are still the same. However, the foreign agent no longer exists in MIPv6. This is because in MIPv6, agent discovery in MIPv4 is replaced by Router Discovery described in [24] to determine its current location. Therefore, the node can operate away from home without any support from the foreign agent. Additionally, all CoA defined for MIPv6 will be collocated CoA, rather than the CoA of the foreign agent. Another significant change in MIPv6 is the security mechanism employed. In MIPv6, IP security (IPSEC) is employed for Binding Updates, unlike MIPv4, which relies on its own security mechanisms. Furthermore, most packets sent to a mobile node when away from home use IP Routing header instead of IP encapsulation, which is necessary for all MIPv4's packets. Finally, the route optimisation mechanism, which is optional in MIPv4, is integrated together with MIPv6.

6.4.2.3 Implementation and Impact of Mobile IP on Satellite Networks

When Mobile IP is used to manage mobility in an integrated system, two levels of mobility management can be envisaged; intra-segment and inter-segment mobility. This is advantageous as the designer of such systems has the option of maintaining segment specific procedures for intra-segment mobility. Therefore, the existing procedures in both segments will remain unchanged.

The application of the Mobile IP procedures is feasible in a satellite system. However, it is also important to highlight the requirements of its implementation in the satellite environment. First, there must be an interconnection of the satellite system to an Internet Core Network, where the Mobile IP protocols run. In addition, the satellite system must be able to make itself known to the other networks in the Internet network. In order to allow interconnection between the satellite system and the Internet network, a FES can be connected to an edge router in the Internet sub-network, therefore allowing IP packet flow. It is also important to ensure that the satellite network itself is capable of supporting IP traffic.

Two other important criteria in utilising Mobile IP in the satellite environment are the assignment of CoA to the mobile node and the mechanisms involved for the communications between the mobile node and the home agent. To address these issues, it is important to ensure that the satellite network is capable of making itself known to other systems in the Internet network. This can be performed by configuring the satellite network as an Autonomous System (AS). An AS is defined as a set of routers under a single technical administration, using an interior gateway protocol and common metrics to route packets within the AS, and using an exterior gateway protocol to route packets to other ASs [19]. Since the Internet is seen as a set of arbitrarily connected AS's, when a satellite network is configured as an AS, it can be seen from the Internet point of view as another Internet subnetwork. AS administrators are responsible for issues such as routing and access policies inside their own AS, however, a consistent interface must be provided in order to communicate with other ASs in the Internet network. Several types of protocol are available for this purpose, such as the Exterior Gateways Protocol (EGP) [14], the Border Gateway Protocol (BGP) [17] and the ISO Inter-Domain Routing Protocol (IDRP). These protocols can be used to exchange reachability information between Internet gateways belonging to the same or different autonomous systems. The current de-facto standard for inter-autonomous routing protocol is the BGP protocol, which was designed based on the older EGP [19].

BGP works by maintaining routing tables and transmitting routing updates to other ASs. In BGP, routers that communicate with other routers via BGP are known as BGP speakers. Therefore, the primary functionality of a BGP speaker is to exchange reachability information, which includes information on autonomous system path with other BGP systems. If a satellite network is configured as an AS, some nodes in the satellite network will have to be elected as the BGP speakers. This can be implemented in the Network Control Centre (NCC), as the NCC is responsible for most of the intelligence of the satellite system. On the other hand, an FES can be configured to act as the point of access to the Internet network. The NCC can also be used to assign IP addresses to mobile nodes in cases where the satellite system is used as a foreign network using DHCP.

Assuming that MIPv6 is used for integrating satellite and terrestrial systems, the need for a foreign agent in the system is not necessary, thus, eliminating the need of implementing a foreign agent in the satellite network. Instead, the mobile node will be responsible for informing its own home agent of its current location. However, if MIPv4 is used in the system, some of the satellite components will have to be upgraded to function as a foreign agent.

6.5 Further Research and Conclusions

A space communication channel differs greatly from that offered by a terrestrial, fixed-network environment. This difference severely affects TCP performance, which has been designed with fixed-network operation in mind. For TCP to work well over satellite, be it in a fixed or mobile environment, further research is required in several key areas.

In [6] a comparison between the different strategies to improve TCP performance over wireless links are thoroughly compared, including link layer extensions and split TCP connections. For actual experiments of TCP performance over geostationary satellite links, the interested reader is referred to [2,1,27]. On the other hand, *ad-hoc* protocols that provide a more efficient replacement of TCP are proposed in [9], while extensions for high performance of currently available TCP versions are explained in [15,16].

Further work needs to be done in order to investigate the effect of TCP over non-geostationary satellites, in particular that of Low Earth Orbit, where variable round trip time and intermittent connectivity are the main channel characteristics. This would be able to show how TCP options such as SACK, PAWS, Window Scaling and Time Stamp Option could help to improve TCP performance over satellite in both geostationary and low earth orbit constellations.

The broadcast nature of satellites can be used for multicast push of HTTP servers at the edges of a satellite network, thus improving cache hit rate and considerably increasing WWW throughput, as can be seen in [10] and [26].

The convergence of fixed and mobile networks opens up many new possible service scenarios. The challenge then is to provide a platform in which TCP/IP can be efficiently delivered in both fixed and mobile environments, the latter being particularly hostile. This is the focus of the European Commission funded COST Action COST 272 'Packet-Oriented Service Delivery via Satellite", which is a four-year follow-on project from COST 253 and COST 252, the latter being concerned with mobile-satellite networks. Further information on the COST 272 project can be found at [7].

References

[1] S. Agnelli and V. Dewhurst, 'LAN Interconnection via ATM Satellite Links for CAD Applications: The UNOM Experiment', *Proceedings of IEEE International Conference on Communications*, 1–3, pp. 931–935, 1996.

[2] M. Allman, C. Hayes, H. Kruse and S. Ostermann, 'TCP Performance Over Satellite Links', *Proceedings of the 5th International Conference on Telecommunication Systems*, pp. 456-462, 1997.

[3] M. Allman, C. Hayes and S. Ostermann, 'An Evaluation of TCP with larger initial windows', *ACM Computer Communication Review*, 28(3), pp. 1–52, 1998.

[4] J. Aracil, D. Morato and M. Izal, 'Analysis of Internet Services in IP over ATM Networks', *IEEE Communications Magazine*, 37(12), pp. 92–97, 1999.

[5] J. Aracil and D. Morato, 'Characterizing Internet Load as a Non-Regular Multiplex of TCP Streams', *Proceedings of IEEE International Conference on Computer Communications and Networks*, pp. 94–99, 2000.

[6] H. Balakrishnan, V.N. Padmanabhan, S. Seshan and R. Katz, 'A comparison of Mechanisims for Improving TCP Performance over Wireless Links', *ACM SIGCOMM*, pp. 256–259, 1996.

[7] COST272 'Packet-Oriented Service Delivery via Satellite' Web Site: http:// www.tesa.prd.fr/cost272/

[8] M.E. Crovella and A. Bestavros, 'Self-Similarity in World Wide Web Traffic: Evidence and Possible Causes', *IEEE/ACM Transactions on Networkingworking*, 5(6), pp. 835–846, 1997.

[9] R. Durst, G. Miller and E. Travis, 'TCP Extension for Space Communication', *Proceedings of ACM MobiComm'97*, pp. 15–26, 1996.

[10] X-Y Hu, P. Rodriguez, and E.W. Biersack, 'Performance Study of Satellite-Linked Web Caches and Filtering Policies', *Proceedings of Networking*, pp. 580–595, 2000.

[11] D.B. Johnson and C. Perkins, 'Mobility Support in IPv6', Internet Engineering Task Force, Internet Draft, draft-ietf-mobileip-ipv6-13.txt, November 17, 2000, Work in Progress.

[12] C. Metz, 'IP-over-satellite: Internet Connectivity Blasts Off', *IEEE Internet Computing*, 4(4), pp. 84–89, 2000.

[13] C. Partridge and T. Shephard, 'TCP/IP Performance over Satellite Links', *IEEE Network*, 11(5), pp. 44–49, 1997.

[14] D.L. Mills, 'Exterior Gateway Protocol Formal Specification', *Internet Engineering Task Force RFC904*, April 1984.

[15] V. Jacobson and R. Braden, 'TCP Extensions for Long-Delay Paths', *Internet Engineering Task Force RFC 1072*, October 1988.

[16] V. Jacobson, R. Braden and D. Borman, 'TCP Extensions for High Performance', *Internet Engineering Task Force RFC 1323*, May 1992.

[17] Y. Rekhter and T. Li, 'A Border Gateway Protocol', *Internet Engineering Task Force RFC 1771*, March 1995.

[18] M. Mathis, J. Mahdavi, S. Floyd and A. Romanow, 'TCP Selective Acknowledgement Options', *Internet Engineering Task Force RFC 2018*, October 1996.

[19] J. Hawkinson, 'Guidelines for creation, selection, and registration of an Autonomous System (AS)', *Internet Engineering Task Force RFC1930*, March 1996.

[20] C. Perkins (Editor), 'IP Mobility Support', *Internet Engineering Task Force RFC 2002*, October 1996.

[21] W. Stevens, 'TCP Slow Start, Congestion Avoidance, Fast Retransmit and Fast Recovery Algorithms', *Internet Engineering Task Force RFC 2001*, January 1997.

[22] R. Droms, 'Dynamic Host Configuration Protocol', *Internet Engineering Task Force RFC 2131*, March 1997.

[23] M. Allman, S. Floyd and C. Partridge, 'Increasing TCP's Initial Window', *Internet Engineering Task Force RFC 2414*, September 1998.

[24] T. Narten, E. Nordmark and W. Simpson, 'Neighbor Discovery for IP version 6', *Internet Engineering Task Force RFC 2461*, December 1998.

[25] S. Deering, and R. Hinden, 'Internet Protocol, version 6 (IPv6) Specification', *Internet Engineering Task Force RFC 2460*, December 1999.

[26] P. Rodriguez and E.W. Biersack, 'Bringing the Web to the Network Edge: Large Caches and Satellite Distribution', *WOSBIS'98 ACM/IEEE MobiCom Workshop on Satellite-based Information Services*, Dallas, TX, USA, October, 1998.

[27] F.J. Ruiz, A. Fernández, C. Miguel, J. Aracil, L. Vidaller and J. Pérez, 'The Picoterminal Network: Portable Communications via Satellite', *Proceedings of the TERENA Joint European Networking Congress '95*, Tel Aviv, Israel, 15–18 May 1995.

[28] W.R. Stevens, *TCP/IP Illustrated. Volume I: The Protocols*, Reading, MA, Addison Wesley, 1994.

[29] K. Thompson, G. Miller and R. Wilder, 'Wide-Area Internet Traffic Patterns and Characteristics', *IEEE Network*, 11(6), pp. 10–23, 1997.

Appendix A

Satellite Constellation Design for Network Interconnection Using Non-Geo Satellites

This appendix discusses satellite constellation design. First the properties of satellite orbits are summarised, then an overview of constellation geometry is provided. This is followed by a discussion on the satellite constellation and architecture of recent broadband non-Geo satellite proposals, including Teledesic, WEST, SkyBridge and GIPSE.

A.1 Satellite Constellation Design

The overall design of a constellation as a discrete autonomous system or private network will have a considerable effect upon the design of its various components. The constellation design should therefore be considered as an early criterion in its own right. The choices made here affect the design of the individual modules due to the large number of interrelated factors.

A.1.1 Orbits

Orbital geometry has a considerable effect on the design of satellite constellation network, and influences satellite coverage and diversity, physical propagation considerations such as power constraints and link budgets. Particularly important, from a networking viewpoint, is the resulting dynamic network topology and round-trip latency and variation. As a result of this, the choice of orbits must be considered carefully in order to be able to characterise the resulting class of satellite network accurately.

We can classify the orbital choices for a constellation into two categories, based on whether the orbits are circular or elliptical.

A.1.1.1 Circular Orbits

Satellites in circular orbits can provide continuous coverage of an area of ground beneath (the 'footprint'), but the area covered by the satellite moves as the satellite moves in its orbit. The altitude of these orbits is selected as a result of physical and geometric considerations, including signal power, time of satellite visibility, coverage area, and avoidance of the Van Allen radiation belts.

Low Earth Orbit (LEO)

At altitudes of typically between 500 and 2000 km, beyond upper atmosphere but below the inner Van Allen belt, a large number of satellites are required to provide simultaneous full-Earth coverage. The actual number of satellites required depends upon the coverage required and upon the frequency bands used; large amounts of frequency reuse across the Earth becomes possible to provide large system capacity. Propagation delay between Earth station and satellite is under 15 ms, depending on their relative locations. LEO satellite constellations include:

1. *Voice*: Iridium, Globalstar
2. *Messaging*: Orbcomm
3. *Proposed for broadband data*: Teledesic, SkyBridge

Medium Earth Orbit (MEO)

At altitudes of around 10,000 km, between the inner and outer Van Allen belts, these orbits can permit full Earth coverage with fewer, larger satellites, with increasing resulting delay. Propagation delay between Earth station and satellite is under 40 ms. MEO satellite constellations include:

- *Voice*: ICO
- *Proposed for broadband data*: Orblink, Hughes Spaceway NGSO, @contact

Geostationary Earth Orbit (GEO)

At an altitude of 35,786 km above the equator, the angular velocity of a satellite in this orbit matches the daily rotation of the Earth's surface, and this orbit has been widely used as a result. Providing coverage of high latitudes (above 75°) is generally not possible, so full Earth coverage cannot be achieved with a geostationary constellation. Propagation delay between Earth station and satellite is a minimum of 0.119 s at the equator but increases with increasing latitude and longitude offset. This leads to the widely-quoted half-second round-trip latency quoted for communications via geostationary satellite. GEO satellites and constellations include:

- *Television broadcast*: Astra etc.
- *Voice*: Inmarsat and proposed single satellites for targeted service areas such as Thuraya, ACeS, APMT
- *Proposed for broadband data*: Hughes Spaceway, Loral Astrolink, MMS WEST, Alenia EuroSkyWay

A.1.1.2 Elliptical Orbits

Elliptical orbits differ from the continuous-but-moving coverage of circular orbits in that coverage is only provided when the satellite is moving very slowly relative to the ground while at apogee, furthest from the Earth's surface, and power requirements in link budgets are

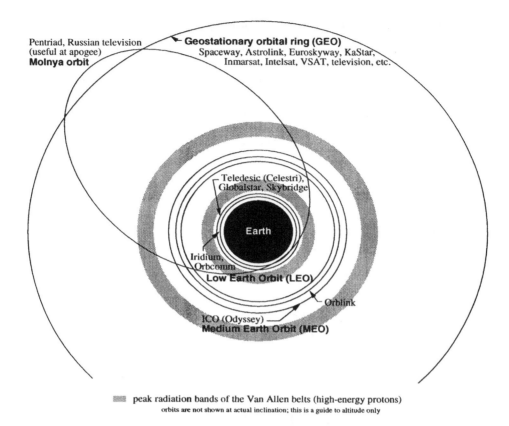

Pentriad, Russian television   **Geostationary orbital ring (GEO)**
(useful at apogee)   Spaceway, Astrolink, Euroskyway, KaStar,
Molnya orbit   Inmarsat, Intelsat, VSAT, television, etc.

Teledesic (Celestri),
Globalstar, Skybridge

Earth

Iridium,
Orbcomm
Low Earth Orbit (LEO)

Orblink

ICO (Odyssey)
Medium Earth Orbit (MEO)

▓ peak radiation bands of the Van Allen belts (high-energy protons)
orbits are not shown at actual inclination; this is a guide to altitude only

Figure A.1   Orbital altitudes for satellite constellations

dimensioned for this distance. When the satellite moves faster from high apogee to low perigee and back, and as its coverage area zooms in size (and other satellites are at apogee providing services in its place) it does not provide service coverage; in fact, its electronics may be shut down while it passes through the Van Allen radiation belts.

These orbits (Figure A.1) are generally at an inclination of 63.4° so that the orbit is quasistationary with respect to the Earth's surface. This high inclination enables coverage of high latitudes, and Russian use of Molnya and Tundra elliptical orbits for satellite television to the high-latitude Russian states is well-known. Elliptical constellations include:

- *Proposed for broadband data*: Virtual GEO and Pentriad. These orbits have apogees beyond the geostationary orbital ring, resulting in larger propagation delays. Virtual GEO also intends to establish intersatellite links between satellites at the apogees of different elliptical orbits.

As elliptical orbits are the exception rather than the general rule, and generally provide carefully-targeted selected coverage, rather than general worldwide coverage, they will not be considered further. The properties of elliptical orbits have been explored extensively by Draim [1].

A.2 An Overview of Constellation Geometry

Simulating a constellation network requires an appreciation of how the satellites move over time, and when handover between satellites occurs. Many satellite constellations are based around the idea of co-rotating planes, slightly offset to provide full overlap, giving *streets of coverage*, as illustrated in Figure A.2.

These co-rotating planes have the same inclination with respect to a reference plane. It would be possible to place these orbital planes at a constant inclination to any reference plane, which does not have to be the equator. However, if we place the orbits with constant inclination to a plane other than the equator, we complicate the ground paths. The varying action of precession due to the oblateness of the Earth will distort the network, making controlling ground coverage and maintaining ISLs via pointing much more complex. As a result, the equator is the only reference plane we consider.

A.2.1 Polar and Near-Polar Star Constellations

The polar orbit has an inclination of exactly 90° and the near-polar orbit constellation has an

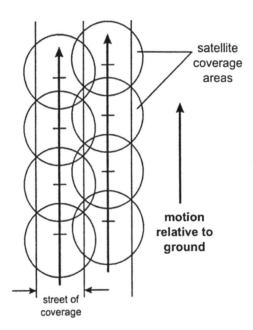

Figure A.2   Streets of coverage

inclination angle near to that, but tailored to the particular requirements of the orbit.

Walker [2,3] explored different types of constellations, often using a streets approach for coverage. Because of his contribution to the field, near-polar constellations with an orbital seam between ascending and descending planes are also named the *Walker star pattern*. This is because all of the orbits cross near the Poles, and if viewed from one of the Poles, the orbital planes intersect to make a star, as illustrated in Figure A.3.

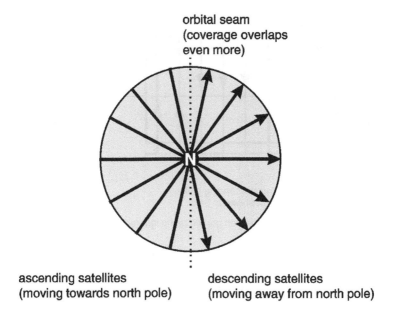

orbital seam
(coverage overlaps
even more)

ascending satellites   descending satellites
(moving towards north pole)   (moving away from north pole)

Figure A.3   Polar view of Walker star pattern

The right ascensions of the ascending nodes of the p orbital planes $\Omega_1..\Omega_p$ are such that they are approximately evenly spaced, with the exception of the two contra-rotating planes at the ring edges, ISLs between the planes may not be supported due to the high relative velocities (twice the orbital velocity) of the satellites moving in opposite directions. The separation between the contra-rotating planes is slightly less than between other planes, to ensure full, overlapping, ground coverage as the streets of coverage move over each other [2].

Figure A.4 ignores the effect on the orbital paths of the Earth's rotation. This effect can be considerable. With the exception of near the Poles, any point on the Earth's surface will see overhead satellites moving at regular intervals from north to south or south to north with a star constellation. As satellites in neighbouring planes are closer to each other at the Poles than at the equator, coverage of the polar constellation is not evenly spread with varying latitude. The equator is the largest separation which must be defined for the coverage and distance between orbits. At the Poles, the overlapping of satellite footprints will cause interference and multiple coverage, requiring some footprints to be disabled. The high relative velocities of satellites travelling in neighbouring planes will make maintaining the ISLs very difficult due to Doppler shift, high tracking rate, and the need to swap neighbours and re-establish links as orbital planes cross.

A.2.2 Delta Constellations

The Delta constellation, also known as the Walker delta, or as a rosette, is discussed in detail in [2]. This is a more general constellation case than the polar case. The orbital planes are inclined with constant inclination δ. The even spacing of the right angles of the ascending nodes $\Omega_1..\Omega_p$ across the full 360° of longitude, means that ascending and descending planes

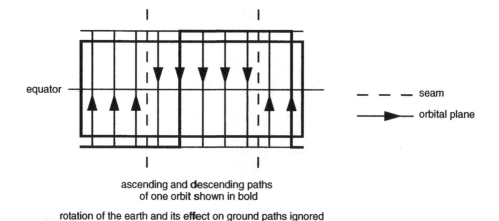

equator

seam

orbital plane

ascending and descending paths
of one orbit shown in bold

rotation of the earth and its effect on ground paths ignored

Figure A.4   Map projection of constellation (stylised)

of satellites continuously overlap (shown in Figure A.5), rather than being completely separate as with the Walker star.

Ballard [4] concentrates on the bounds of multiple satellite visibility by interleaving low-
inclination multiple planes containing few satellites and using careful phasing to fill in the
gaps between satellite footprints in the same plane. Ballard calls the delta constellation an
inclined rosette constellation and does not use the 'street of coverage' approach (required for
near-polar constellations with near-parallel planes) that is assumed by Rider and detailed in
[5]. Although interleaving and careful phase alignment of planes for coverage is adopted by

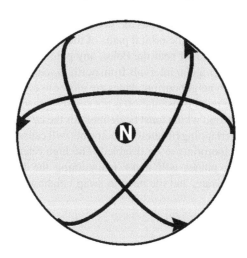

no orbital seam;
ascending and descending satellites overlap

Figure A.5   Simplest Walker delta constellation

SkyBridge to decrease the number of satellites required, the inclined orbits place more severe constraints on inter-satellite networking with intersatellite links.

There is no coverage above a certain latitude depending upon the value of δ; inclined rosette constellations generally neglect polar coverage.

A.2.3 Notation

Constellations are usually described in one of two forms in the literature:

Walker notation: *N/P/p* where *N* is the number of satellites per plane; *P* the number of planes *P*; and *p* the number of distinct phases of planes to control spacing offsets in planes.

Thus, a Teledesic Boeing design approximation could be described as 24/12/2.

Ballard notation: *(NP,P,m)* where *NP* is the total number of satellites; *P* the number of planes; and *m* the harmonic factor m describing phasing between planes.

A SkyBridge design approximation could be described as two subconstellations, both (40,8,2), offset from one another.

The Walker notation is more commonly seen, although the Ballard notation can more accurately describe possible offsets between planes, especially when *m* is a fractional. Note that the Ballard notation is assumed to describe a rosette where the ascending nodes are spaced over 360°, while in Walker notation this is unspecified and unclear.

A.3 Network Topology

A.3.1 Primarily Ground-Based Networks

The topology of a ground-based network, where the satellites are only used to provide last-hop connectivity, is entirely arbitrary, and governed primarily by financial considerations. It is likely that all satellite telemetry, tracking and control (TT and C) ground stations will be networked, to share information about the state of the constellation, but beyond that, there are a large number of networking possibilities and a number of ways the constellation can be integrated with existing terrestrial networks.

As a result of this, the design of the terrestrial network component of a ground-based constellation like SkyBridge must be explicitly determined.

A.3.2 Primarily Space-Based Networks

Primarily space-based networks have a slightly more predictable topology, as a result of their use of ISLs and the constraints placed on ISL use by orbital geometry. (Of course, these space-based networks can be complemented by ground networks as briefly discussed above, opening up the design space).

A.4 Intersatellite Links

It is possible to design a satellite constellation network as a primarily ground-based network where an overhead satellite simply provides the last hop, where radio or laser ISLs provide direct connectivity between satellites, or as a combination of the two approaches providing redundancy.

The ISL approach, where satellites communicate directly with each other by line of sight, can decrease Earth–space traffic across the limited air frequencies by removing the need for multiple Earth–space hops. However, it requires more sophisticated and complex processing/switching/routing onboard satellite to support the ISLs. This allows completion of communications in regions where the locally-overhead satellite cannot see a ground gateway station, unlike simpler 'bent-pipe' frequency amplifying/shifting satellites which act as simple transponders.

For circular orbits, fixed fore and aft ISL equipment to communicate with satellites in the same plane is possible. However, interplane ISL is impossible since the direction and the length the line-of-sight paths between satellites on different planes change as the satellites separate and converge between orbit crossings, giving rise to:

- High relative velocities between the satellites;
- Tracking control problems as antennas must slew around;
- Doppler shift.

In elliptical orbits, a satellite would see the relative positions of satellites 'ahead' and 'behind' appear to rise or fall considerably throughout the orbit. Controlled pointing of the fore and aft intraplane link antennas would be required to compensate for this, whereas interplane crosslinks between quasi-stationary apogees can be easier to maintain (the Virtual GEO design).

Of the proposed broadband data constellation to be discussed in A.7, Teledesic, is an example of a space-based network using ISLs, while SkyBridge is an example of a ground-based network without ISLs. Both proposals planned for similar orbital altitudes at about 1400 km. These two different satellite constellations provide a perfect scenario for comparing the performances of the two different approaches.

A.5 Design Choices

The number of possible physical design choices for a constellation are considerable, and not easily summarised. From a networking perspective and at a macroscopic level, the following constellation choices can be considered:

- Space-based network with ISLs vs. ground-based network without ISLs;
- Star constellation without overlapping ascending and descending planes vs. seamless delta constellation with overlapping planes.

In both cases, the former represents a Teledesic-like constellation whereas the latter the SkyBridge-like constellation. Although the ground network topologies are undefined, comparison can be made by considering identical users with identical requirements in identical locations. Figures A.6 and A.7 show the protocol stacks for the two different approaches.

The total number of satellites required in the constellation depends on the orbital types, the design methodology, and the minimum elevation angle which is required of the system. Another important factor is the cost and the capability of launching a satellite. It is more expensive to launch satellites into higher orbits than in lower orbits. Besides less payload

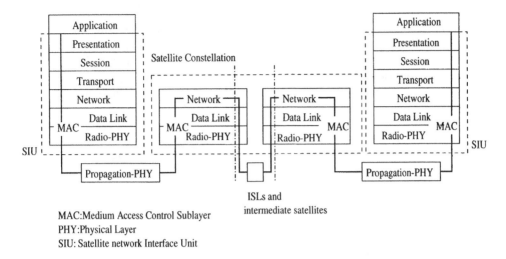

MAC:Medium Access Control Sublayer
PHY:Physical Layer
SIU: Satellite network Interface Unit

Figure A.6   ISL routing approach

mass can be launched into higher orbits. The satellite payload architecture, as described in Chapter 4, also impacts on the cost of the satellite and the interworking complexity.

A.6 Coverage, Availability and Diversity

It is not necessary for a satellite system to provide a true global coverage. For example, the demand for network interconnection at the poles and in the oceans would be minimal. Satellite constellation and capabilities should be tailored to the target coverage area. With a non-geostationary system, this concept might be difficult to achieve, however desirable.

The minimum elevation angle of the satellite constellation will determine the probability of shadowing and the depth of the shadowing attenuation. The channel should be stable to avoid interruption. This might imply higher satellites or more satellites.

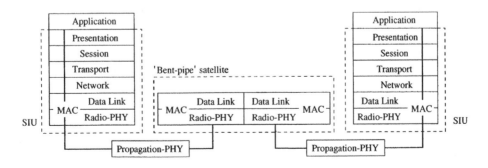

MAC:Medium Access Control Sublayer
PHY:Physical Layer
SIU:Satellite network Interface Unit

Figure A.7   Relay 'bent-pipe' satellite approach

Earth-to-space diversity is the use of more than one satellite at the same time for communication. This allows an improvement in physical availability, by decreasing the impact of shadowing (buildings obstructing the path between the ground terminal and satellite) and providing redundancy at the physical or data-link level. Diversity is also exploited for soft handovers. An overview of Earth-to-space diversity can be found in [6].

The presence of two satellites simultaneously on view to the user can be beneficial in a shadowing environment. Signals from two satellites could be combined in order to combat deep fades due to multipath and other forms of channel degradation. Satellite diversity implies that a great deal of frequency co-ordination must be performed to ensure that the satellites serving the same geographic area do not interfere with each other. Diversity could also imply that a larger number of satellites is required compared with a low-diversity system, although in some designs diversity is inherent and may not increase the number of satellites required. In this case, high diversity levels may be limited in latitude and not apply to all parts of the globe.

Earth-to-space diversity can be applied to various layers of the protocol stack. Physical diversity can be used in Rosette constellation. One such example is Globalstar which makes use of CDMA and recombination of signals across multiple satellite transponders. It can also be applied at the data link layer via TDMA management. At present, code diversity and network-layer diversity have not been planned for use in commercial constellations.

A.8 Proposed Non-Geo Broadband Satellite Systems

A.8.1 Teledesic

A.8.1.1 Satellite Constellation

Teledesic [7] is a broadband Ka-band satellite constellation aimed at providing global services to a mix of users, from individual users to large company networks. The original Teledesic design, announced in 1994, required 840 active satellites in 21 sun-synchronous near-polar orbital planes of 40 satellites per plane. In 1997, Boeing partnered with Teledesic and committed to producing the satellites. A constellation redesign for 288 larger active satellites in the constellation, in 12 near-polar planes of 24 satellites per plane has since been announced, but full technical details of the constellation have not been made available (Teledesic later allied with Motorola, who are understood to be undertaking a redesign. Details of the Motorola design are not known).

Like the original design, the redesign uses frequencies in the Ka-band on uplinks and downlinks allocated by the FCC and agreed at WRC-95 and -97. Downlinks operate between 18.8 and 19.3 GHz, and uplinks operate between 28.6 and 29.1 GHz. Atmospheric attenuation and rain fade at these frequencies are countered by specifying a high minimum elevation angle for ground stations of 40° to give 99.9% availability globally, although outages during tropical monsoons may occur. Keeping the minimum elevation angle while decreasing the number of satellites in the constellation during the redesign resulted in a change of constellation altitude from 700 to 1350 km.

The original design discussed channels specified with a granularity of 16 kbps, rising from 16 kbps for the smallest mobile terminals up to E1 (2 Mb/s) and higher in certain circumstances. For the redesign, Teledesic expect that most customers will have two-way, asymmetric, connections that provide up to 64 Mbps on the downlink and up to 2 Mbps on the uplink. Symmetric broadband terminals will offer 64 Mbps of two-way capacity. However, little is known about the details of implementation.

The Teledesic company foresees a crossover shift of services in the near future. Those services which are currently delivered on satellite will cross over to fixed fibre (TV to cable) and vice-versa (Internet to satellite). This is a variation of the widely-cited 'Negroponte switch'.

A.8.1.2 Architecture of Satellites

On-Board Switching

The Teledesic satellites employ a proprietary non-blocking fast packet switch using proprietary switching protocols and algorithms rather than ATM. Very little is known publicly about this. The combination of switch and protocols are said to produce a intelligent network equivalent to layer 2 of the OSI stack, which will be transparent to other protocols travelling over it, be they IP, ATM, or anything else.

Teledesic provides a potential instance of a multimedia satellite system, although for reasons set out earlier it could be more efficient to ground the traffic as early as possible, as in a MEO system, and route it through the terrestrial network in order to reduce delay. This is a trade-off between the switching complexities of the satellite networks and of the ground networks, and the speed of light in vacuum and in fibre.

Non-blocking switches are apparently still planned for the 288-satellite constellation. It is said that Teledesic will support several priorities of traffic, and that low-priority traffic may experience queuing delays over congested inter-satellite links. How this QoS differentiation will be implemented, and details of implementation support for other networking issues such as multicast, are unknown.

ISLs

So far, non-geostationary ISLs have been proposed for some LEO constellations, in particular, Iridium (of the voice LEOs), Teledesic, and Celestri. In these systems, the ISLs are maintained between adjacent satellites in the same orbital plane and also with nearby satellites in adjacent planes.

Coverage of the Earth by any single LEO satellite, is a small fraction of the Earth's surface. In LEO systems without ISLs, each satellite needs to be able to 'see' a ground station, otherwise it will be unable to route its traffic to the ground. The number of ground stations required for full satellite connectivity will be large and will increase the total cost of the system. However, ground stations need only be implemented where the potential market and resulting revenue justifies the cost. Areas of the world which do not have ground stations will not have access to the satellite services. ISLs would reduce the number of ground stations needed by routing traffic through the satellite network to the nearest or most appropriate ground station.

ISLs in the Teledesic system will use on-board packet switching for a true 'computer telecommunications network' functionality. Up to eight simultaneous ISLs were specified for each satellite in the original 840-active-satellite design, covering both in-plane and inter-plane (cross-link) ISLs. This connectivity is still possible in the 288-satellite redesign. However, the original design called for phased-array RF antennas in the ISL payloads.

The increasing maturity of optical ISLs and the demand for ISL capacity above 1 Gbps has resulted in a switch to optical ISLs for the redesign.

A.8.2 SkyBridge

A.8.2.1 Satellite Constellation

The SkyBridge [8] constellation is a Ku-band broadband constellation that consists of two identical, slightly offset, inclined rosettes giving full ground coverage above a minimum elevation angle of ten degrees between $\pm$ 68° of latitude, although due to a lack of ISLs that coverage is only effective when a user is in the same satellite footprint as a gateway station.

Each rosette previously consisted of eight orbital planes; now it is 10 orbital planes of four active satellites per plane. SkyBridge is intended to share frequencies with existing GEO satellites in Ku-band, and the two overlapping offset constellations are intended to offer diversity. When one satellite in a co-orbiting pair causes interference with a GEO satellite, the alternate satellite of the pair is used. It has not yet been ascertained how well this suggested design will work in practice.

A.8.2.2 Architecture of Satellites

On-Board Switching

The SkyBridge design does not utilise on-board processing or on-board switching. As SkyBridge is a LEO constellation at approximately the same altitude as Teledesic, this is an interesting contrast. Instead, the 45 spotbeams on each satellite are to be handled via a set of phased-array antennas interconnected via transponders providing channel filtering and frequency conversion. This means that a double-hop is necessary between users on the same satellite.

Handovers are pre-computed by the gateway stations, and each user terminal is instructed on the handover procedure by its associated gateway when the handover is due to take place.

ISLs

SkyBridge does not utilise intersatellite links – this not surprising, as OBP and baseband switching are absent and the satellites are simple frequency-shifting-and-amplifying 'bent pipes'. Routing for ISLs would therefore not be possible, and an ISL payload would be redundant.

This lack of ISLs effectively restricts SkyBridge's network coverage to areas of land masses containing ground stations to sink and source traffic to and from the terrestrial network, and results in a 'double hop' for communications between two SkyBridge users. However, the lack of ISL routing, makes the SkyBridge satellite a transparent path between users and the local gateway station, where ATM interfaces to the ground networks are expected. SkyBridge can be considered as a transparent extension of the ground network for multicast applications.

User Terminal Types

Users are perceived as falling into one of three groups: residential users; professional (corporate) users; or internode (large corporate) users, and are summarised in Table A.1.

The Professional and Internode designs are of interest from a corporate LAN viewpoint – programmed-mode mechanically-steered antennas are proposed for their user terminals.

Table A.1   User Terminal types for SkyBridge

User type	Residential	Professional	Internode
Size of outdoor terminal (m)	Under 0.5	Over 0.6	Over 0.6
Uplink data rate (Mbps)	2.56	Multiples of 2.56	Multiples of 2.56
Downlink bit rate (Mbps)	20.48	Multiples of 20.48	Multiples of 20.48
Duplexing technique	Time	Frequency	Frequency
Availability (%)	Greater than 99	Greater than 99.5	Greater than 99.9

A.8.3 WEST

A.8.3.1 Satellite Constellation

WEST [9] is an example of a hybrid constellation which uses more than one type of orbit plane within the same system. WEST proposes to augment the coverage of 12 GEO satellites with a constellation of 9 MEO satellites, which are placed in orbits based on a design presented to the COST227 group [10].

Deployment of the full system is in several phases. Initially, GEO satellites will be launched and targeted at those areas of the globe which offer financial returns on the initial investment, i.e. developed countries and large densely-populated developing countries. The returns from this part of the system will be used to help finance further second phase GEOs and the MEO system. This gradual build-up of coverage and capacity is one of the marketing strong points which the WEST proponents MMS are highlighting. The exact orbit locations of the GEO satellites are unknown at this time, but it is known that WEST targets the IP and ATM market using DVB-S on the forward link.

The orbit design of the MEO constellation ensures that satellite coverage will be concentrated on three main paths from north to south (and vice versa) over the continental land masses. These three areas are defined as: (1) North and South America; (2) Europe and Africa; (3) Asia and Australia.

Ocean coverage is not targeted by this MEO system. This coverage targeting is achieved by the careful choice of altitude, inclination angle, satellite phasing and RAAN. All satellites in the MEO constellation are intended to follow exactly the same subsatellite track. The MEO satellites have a period of 8 h, which places their altitude at a constant 13,890 km. This causes a latency slightly greater than that of the previously proposed MEO system, such as ICO, although it is not clear whether this extra delay will be significant. Equally important, is the minimum elevation angle of the MEO satellites. It is clear that the MEO system can only be used with the GEO system for providing the high service rates intended. Without the accompanying GEO coverage, a MEO system such as this would have to offer very low elevation angles for global or near global coverage.

The WEST system also proposes to split the service provision between the two orbit types. The GEO satellites will offer fixed broadband, delay insensitive services. The MEO satellite constellation will be able to increase the capacity of the system, overlapping with some of the areas covered by the GEO satellites and offering the same services, while extending the service range to offer real-time delay sensitive services.

A.8.3.2 Architecture of Satellites

ISLs
Both satellite types will use optical ISLs. GEO satellites will each have two ISL links to adjacent GEO satellites; whereas MEO satellites will be able to communicate with up to four other MEO satellites. Two of these satellites will be those that precede and follow the subject satellite along the subsatellite track. It is not clear where the other two satellites will be placed, although the distances between these satellites will be large. The GEO ISLs will carry 1.2 Gbps traffic and the MEO ISLs each will carry 622 Mbps.

On-Board Switching
The WEST satellites will switch ATM packets, using statistical multiplexing of different traffic types. The switch will provide full autonomous routing connectivity between spot-beams on the same satellite, including multicast operation. The maximum satellite switch capacity will be 8.5 Gbps. The system will re-use ATM protocols on-board the satellite as well as in the ground segment, enabling the provision of bandwidth on demand to users.

A.8.4 GIPSE

A.8.4.1 Satellite Constellation
The GIPSE constellation is comprised of 24 MEO satellites equally distributed within six circular orbit planes, inclined at 45° to the equatorial plane. The altitude of the orbit is 10,350 km, which gives an orbit period of approximately 2 h. If such a constellation were to be deployed it is likely that one in-orbit spare per plane would be launched also.

This particular constellation configuration was designed to minimise the number of satellites deployed while offering a high minimum elevation angle to the users in the designated service area of ±65° latitude. The elevation angle performance is such that in the worst case, users can always see a satellite above 40°, which rises to a best case minimum elevation angle of approximately 53° at 30° latitude. User terminals in this system will be restricted in uplink eirp due to safety considerations and downlink G/T will be restricted due to antenna size and cost. A stable channel which results from high satellite elevation angles will allow a lower link margin for a high availability. An inclined LEO constellation offering comparable minimum elevation angle performance in the same service area would require in excess of 160 satellites, a polar or near-polar constellation will require even more.

The latency which results from a MEO system will naturally be much higher than for a LEO system, although the high minimum elevation angle to the user reduces the latency in the service link compared with a low elevation angle MEO system (ICO). Latency jitter will be low, due to the low frequency of handover. The maximum latency variation which might occur during inter-satellite handover depends on the changes in path distance between both user and satellite, and ground station and satellite. For the user–satellite path, the difference in propagation distance between the subsatellite point and the edge of coverage is in the order of

1400 km (4 ms). For the ground station–satellite path the distance is approximately 4500 km (15 ms). In theory the maximum change in latency during handover could be 19 ms, although this magnitude of differential is unlikely in a real handover scenario.

A.8.4.2 Architecture of Satellites

ISLs

The GIPSE constellation will most likely not encounter the problems of ground station visibility which apply to the previously mentioned LEO systems. Seven ground stations, distributed globally, can give full connectivity to all satellites over the target service area above elevation angles of 5°. The inclined orbit planes, while providing good satellite diversity to users, increase the co-plane inter-satellite dynamics, to a lesser extent in MEO satellites than in LEO. As a result, ISL payloads would need to track adjacent satellites over a large angular range and over much larger distances. Although GEO ISLs may operate over larger distances, the inter-satellite dynamics are much reduced. If one satellite were to be used to ground the traffic of another, the increase in payload size might be prohibitive as each satellite in this constellation will carry a large amount of traffic compared with those in a LEO system.

As mentioned above, ISLs in a MEO system will need to cover large distances in order to link adjacent satellites. Regardless of the hardware techniques required to achieve this, the gains in latency are less obvious than in LEO. A single ISL hop in a MEO system could add another 50 ms or more in latency, if it were used to route traffic to an adjacent ground station which was not available to the primary satellite. There could be benefits in some GIPSE user–GIPSE user links, but in such a highly integrated system, where fixed and GIPSE services are similar, it is unlikely that this will be a significant proportion of the overall traffic.

On-Board Switching

The GIPSE system is intended to be highly integrated with the future B-ISDN and therefore the satellites will carry an ATM derived switch. The ATM cell headers will be slightly different in this system from that standardised for terrestrial fibre-based systems, enabling higher efficiency in the Satellite-UNI or air interface. Some of the call control functions associated with a future mobile broadband network will be based on the ground to simplify the satellite segment, although time-critical functions such as congestion control and inter-beam handover will be handled autonomously by the satellite.

References

[1] J. Draim, 'Satellite continuous coverage constellations', US patent 4809935, 'Elliptical satellite system which emulates the characteristics of geosynchronous satellites', US patent 58745206 and others.

[2] J.G. Walker, 'Some circular orbit patterns providing continuous whole Earth coverage' *Journal of the British Interplanetary Society*, Vol. 24, pp. 369–384, 1971.

[3] J.G. Walker, 'Satellite constellations', *Journal of the British Interplanetary Society*, Vol. 37, pp. 559–571, 1984.

[4] A.H. Ballard, 'Rosette constellations of Earth satellites', *IEEE Transactions on Aerospace and Electronic Systems*, Vol. 16, No. 5, 1980.

[5] L. Rider, 'Optimised polar orbit constellations for redundant Earth coverage', *Journal of the Astronautical Sciences*, Vol. 33, pp. 147–161, 1985.

[6] T.E. Wisløff, Dual satellite path diversity in non-geostationary satellite systems, Dr. Ing. Thesis, the Norwegian University of Science and Technology, Department of Telematics, 1996.

[7] M.A. Sturza, 'Architecture of the Teledesic satellite system', *Proceedings of the International Mobile Satellite Conference*, pp. 212–218, 1995.

[8] P.L. Spector, J.H. Olson, et al., FCC filing, 'Application of SkyBridge LLC for authority to launch and operate the SkyBridge system', SkyBridge LLC, February 1997.

[9] B. Le Stradic, et al., 'WEST: this evolutive Ka-band satellite system provides Europe with a cost-effective alternative to the numerous US contenders for multimedia services', Fourth European Conference on Satellite Communications (ECSC-4), Rome, Italy, pp. 28–35, 1997.

[10] G. Pennoni, 'JOCOS 6+1 Satellites to evolve towards Universal Personal Telecommunications', COST227 TD(94)-18.

Appendix B

List of temporary documents[a]

COST253 TD(98) 001	'An overview of Source modelling', T. Ors, Z. Sun, B.G. Evans, University of Surrey
COST253 TD(98) 002	'On Internet Traffic Self - Similarity', Javier Aracil, University Publica de Navarra
COST253 TD(98) 003	'Management Software for non-geo broadband satellite systems', Technologies de Telecomminicationes y de la Informacion
COST253 TD(98) 004	'An overview of future satellite communication options for LAN interconnection', Tolga Ors, Tony Sammunt, Lloyd Wood, Barry Evans, University of Surrey
COST253 TD(98) 005	'Cost 253 topics where Telenor may contribute', Vendela Paxal, Telenor
COST253 TD(98) 006	'Routing in Low Earth Orbit Satellites',Laurent Franck, University of Brussels
COST253 TD(98) 007a	'Multicast in satellite communications', L. Wood, University of Surrey (paper)
COST253 TD(98) 007b	'Multicast in satellite constellations', L. Wood, University of Surrey (transparencies)
COST253 TD(98) 008	'Congestion control mechanisms in LEO constellation using ISLs and based on ATM', Fairouz Dabbarh, University of Brussels
COST253 TD(98) 009	'On the performance of ALOHA Channels under Self-Similar Input', Javier Aracil, University of Navarra
COST253 TD(98) 010	'Proposal for a simulator architecture', Erina Ferro, CNUCE-CNR
COST253 TD(98) 011	'Interconnecting LANs through satellites-security related issues', Dr Denis Trcek (J. Stefan Institute)
COST253 TD(98) 012	'Traffic and Routing simulator for LEO constellations', E. Papapetrou, I. Gragopulos, F.N. Pavlidou (Aristotle University of Thessaloniki)
COST253 TD(98) 013a	'Traffic matrix estimation in ATM networks, Methods and solutions', Garcia Gutierrez Alberto E., Hackbarth Planeta Klaus D. (Universtity of Cantabria)

COST253 TD(98) 013b   'Traffic matrix estimation in ATM networks, Methods and solutions' (presentation), Garcia Gutierrez Alberto E., Hackbarth Planeta Klaus D. (Universtity of Cantabria)

COST253 TD(98) 014   'A framework for MAC protocols analysis to provide internet services on LEO satellites', J. Villadangos (University of Navarra)

COST253 TD(98) 015   'LeoSim - a routing simulator for LEO's', L. Franck (Brussels University)

COST253 TD(98) 016a   'Authentication protocols to secure satellite ATM networks', Haitham Cruickshank (University of Surrey)

COST253 TD(98) 016b   'Authentication protocols for satellite ATM networks' (presentation), Z. Sun, Haitham Cruickshank (University of Surrey)

COST253 TD(98) 017a   'Security systems design and implementation examples in ATM networks', Haitham Cruickshank (University of Surrey)

COST253 TD(98) 017b   'ATM network security systems examples', Z. Sun, Haitham Cruickshank (University of Surrey) (presentation)

COST253 TD(98) 018   'COSIM – Concept and Architecture', Marco Annoni (CSELT)

COST253 TD(98) 019a   'Simulation of the satellite constellation and the ground stations that access it' E. Ferro (CNUCE/C.N.R.), L. Franck (University of Brussels)

COST253 TD(98) 019b   'On simulation for COST 253 action' E. Ferro (CNUCE/C.N.R.), L. Franck (University of Brussels)

COST253 TD(98) 020   'Resource allocation and connection admission control in satellite networks', H. Koraitim, S. Tohme (ENST)

COST253 TD(99)001a   'GALILEO: A simulator of the satellite constellation and the ground stations that access it', E. Ferro (CNUCE/CNR), L. Franck (University of Brussels)

COST253 TD(99)001b   'GALILEO', E. Ferro (CNUCE/CNR), L. Franck (University of Brussels)

COST253 TD(99)002   'Performance evaluation of LEO satellite constellations with inter-satellite links under self-similar and Poisson traffic', E. Papapetrou, I. Gragopoulos, F.-N. Pavlidou (Aristotle University of Thessaloniki)

COST253 TD(99)003   'Performance study of adaptive routing algorithms for LEO satellite constellations under self-similar and Poisson traffic', I. Grapopoulos, E. Papapetrou, F.-N. Pavlidou (Aristotle University of Thessaloniki)

COST253 TD(99)004a   'Routing in Low Earth Orbit satellite constellations with inter-satellite links', L. Franck (University of Brussels)

COST253 TD(99)004b   'Routing in Low Earth Orbit satellite constellations with inter-satellite links' (presentation), L. Franck (University of Brussels)

COST253 TD(99)005	'Doppler Frequency Shift Correction for Mobile satellites', V. Paxal, B. Ficini (Telenor R&D)
COST253 TD(99)006a	'Reliability modelling for satellite constellation systems', M. Annoni (CSELT)
COST253 TD(99)006b	'Reliability modelling for satellite constellation systems', M. Annoni (CSELT). (presentation)
COST253 TD(99)007	'Evaluation of W-CDMA parameters for TCP/IP connectivity over LEO satellite networks', J. Aracil, J. Villandangos (University Publica de Navarra
COST253 TD(99)008	'An extended analytical model of guranted handover in satellite fixed cell systems', Santiago Jaramillo (ENST/France)
COST253 TD(99)009	'UPNA-CNUCE short scientific mission', Javier Aracil (Public University of Navarra)
COST253 TD(99)010a	'GaliLEO Progress Report', N. Celandroni, E. Ferro, F. Potorti' (CNUCE/CNR), L. Franck (ENST/France)
COST253 TD(99)010b	'GaliLEO Progress Report', N. Celandroni, E. Ferro, F. Potorti' (CNUCE/CNR), L. Franck (ENST/France) (presentation)
COST253 TD(99)011	'Review of Research in WG1', Niovi Pavlidou (Aristotle University of Thessaloniki)
COST253 TD(99)012	'An Overview of LeoSim through the study of three routing information distribution schemes', L. Franck, G. Maral (ENST/France)
COST253 TD(99)013a	'Simulation Tool for Performance Evaluation of Adaptive Routing Algorithm in ISL Networks, M. Mohorcic, A. Svigelj, G. Kandus (Jozef Stefan Institute/Slovenia)
COST253 99)013b	'Simulation Tool for Performance Evaluation of Adaptive Routing Algorithm in ISL Networks', M. Mohorcic, A. Svigelj, G. Kandus (Jozef Stefan Institute/Slovenia) (presentation)
COST253 TD(99)014a	'Overview of Candidate Coding and Modulation techniques for Packet Switched Traffic over Satellite', Vandela Paxel (Telenor/Norway)
COST253 TD(99)014b	'Overview of Candidate Coding and Modulation techniques for Packet Switched Traffic over Satellite' (presentation), Vandela Paxel (Telenor/Norway)
COST253 TD(99)015a	'Variable rate N-MSK modulation technique for non-terrestrial communication systems', T. Javornik, G. Kandus (Jozef Stefan Institute/Slovenia)
COST253 TD(99)015b	'Variable rate N-MSK modulation technique for non-terrestrial communication systems', T. Javornik, G. Kandus (Jozef Stefan Institute/Slovenia) (presentation)
COST253 TD(99)016	'Broadband Satellite Networks', Tolga Ors (Nortel Networks)

COST253 TD(00)001	'Evaluation of two approaches for implementing routing algorithm in LEO constellation' (presentation), I. El Khamlichi (University of Brussels), L. Franck (ENST)
COST253 TD(00)002	'Short Term Scientific Mission in LMU (presentation)' M. Mohorcic, Institut Josef Stefan
COST253 TD(00)003	'Galileo – A Simulator for Everyone (presentation)', L. Franck, ENST
COST253 TD(00)004	'Simulation Tool for Performance Evaluation of Adaptive Routing Algorithms in ISL Networks'(presentation), M. Mohorcic, Institut Josef Stefan
COST253 TD(00)005a	'Moments of the total interference power in a LEO Satellite System' (paper), M. Remiche, University of Brussels
COST253 TD(00)005b	'Marked IPhP3, an illustration: moments of the total interference power in a CDMA based LEO satellite Communication System' (presentation), M. Remiche, University of Brussels
COST253 TD(00)006a	'An Overview of LEOSim through the study of three routing information', L. Franck, ENST
COST253 TD(00)006b	'Candidate routing algorithms for LEOs', L. Franck, ENST
COST253 TD(00)007	'Performance evaluation of adaptive ISL routing in uniform traffic load condition', M. Mohorcic, Institut Jozef Stefan
COST253 TD(00)008	'Signalling for routing in LEO Constellations', L. Franck, ENST

[a] TD(XX)NNN. Where TD, temporary document; XX, two last digits of the year; NNN, number of document.

List of external documents[a]

COST253 ED(98) 001	'Skybridge', by Michel Cohen, Skybridge
COST253 ED(98) 002	'Analysis and Modelling of Traffic in Modern Data Communications Networks, G. Babic, B. Vandalore, R. Jain, Ohio State University, Department of Computer and Information Science, http://www.cis.ohio-state.edu/~jain/
COST253 ED(98) 003	'Mobile VCE. General information and Progress on Service work programme', T. Ors (University of Surrey)
COST253 ED(98) 004	'Multimedia services modelling approach', S. Ammassari (University of Bradford)
COST253 ED(99)001	'Planning procedure for the UTRA FDD mode in STORMS', R. Menolascino (CSELT)

[a] ED(XX)NNN. Where ED, external document; XX, two last digits of the year; NNN, number of document.

Appendix C

List of Publications

[1] M. Annoni and S. Bizzarri 'An Integrated Modelling Methodology for the Simulation of the Life Cycle of Satellite Constellation', *Joint International Workshop*, Supaero, Toulouse, France, May 1999.

[2] J. Aracil and L. Muñoz, 'Performance of Aloha channels under self-similar input', *IEEE Electronics Letters*, Vol. 33, No. 8, 1997.

[3] J. Aracil, D. Morató and M. Izal, 'Use of CBR for IP over ATM', *Proceedings of the SPIE Voice, Video and Data Communications Conference*, Dallas, TX, November 1997.

[4] J. Aracil, D. Morato and M. Izal, 'On the Combined Effect of Self-Similarity and Flow Control in Quality of Service for Transactional Internet Services', In Korner and Nilsson (Editors), *Performance of Information and Communications Systems*, London, Chapman & Hall, 1998.

[5] J. Aracil, D. Morato and M. Izal, 'Analysis of Internet Services in IP over ATM Networks', *IEEE Communications Magazine*, Vol. 37, No. 12, 1999.

[6] N. Celandroni, E. Ferro, F. Potorti and L. Franck 'A Simulation Tool For Traffic on Leo Satellite Constellations' *Joint International Workshop*, Supaero, Toulouse, France, May 1999.

[7] I. Gragopoulos, E. Papapetrou and F.-N. Pavlidou, 'Performance Study of Routing Algorithms for LEO Satellite Constellations', *Proceedings of ICT '99*, Korea, pp. 144–148, 1999.

[8] I. El Khamlichi and L. Franck, 'Study of two policies for implementing routing algorithms in satellite constellations', *18th AIAA International Satellite Communications Systems Conference*, Oakland, April 2000.

[9] L. Franck and G. Maral, 'Candidate algorithms for routing in a network of inter-satellite links', *18th AIAA International Satellite Communications Systems Conference*, Oakland, pp. 805–815, April 2000.

[10] I. Gragopoulos, E. Papapetrou and F.-N. Pavlidou, 'Performance Study of Routing Algorithms for LEO Satellite Constellations', *Proceedings of ICT '99*, Korea, pp. 144–148, 1999.

[11] I. Gragopoulos, E. Papappetrou and F.-N. Pavlidou, 'Performance study of adaptive routing algorithms for LEO satellite constellations under self-similar and Poisson traffic', *Space Communications*, in press, 2001.

[12] T. Javornik and G. Kandus. 'Variable rate CPFSK modulation technique', In M. Ruggieri (Editor), *Mobile and Personal Satellite Communications 3. Proceedings of the Third European Workshop on Mobile/Personal Satcoms (EMPS'98)*, Venice, Italy. London: Springer-Verlag, pp. 376–388, 1998.

[13] L. Vidaller, J. Aracil, F.J. Ruiz, A. Ruiz and J. Perez. 'Multiple Access Schemes for ATM over Satellite (abstract)'. In P.G. Sterben, J. Touch (Editors), 'Report on the 1995 IEEE Gigabit Networking Workshop', *IEEE Network Magazine*, Vol. 9, No. 4, 1995.

[14] H. Koraitim and S. Tohmé. 'A Movable-Boundary Integrated CBR/Bursty-Data Traffic in Star-Configured VSAT Satellite Networks', *IEEE Symposium on Computers and Communications, ISCC '97*, Alexandria, Egypt, June 97.

[15] H. Koraitim and S. Tohmé. 'Movable Boundary Policies for Resource Allocation in Satellite Links', *ICCC'97 Conference*, Cannes, France, November 1997.

[16] H. Koraitim and S. Tohmé. 'The impact of the Threshold Value On the Performance of the DMBS Allocation Scheme', *Networld and Interop98 Engineers Conference*, Las Vegas, USA, May 1998.

[17] H. Koraitim, S. Tohmé, M. Berrada and A. Brajal. 'Performance of multiple access protocols in geo-stationary satellite systems', *IFIP HPN'98 Conference*, Vienna, Austria, September 1998. Dordrecht, Kluwer 1998.

[18] H. Koraitim and S. Tohmé. 'Resource allocation and connection admission control in satellite networks', *IEEE Journal Selected Areas in Communication*, Vol. 17, No. 2, 1999.

[19] H. Koraitim and S. Tohmé. 'A multiple access protocol for packet satellite networks', *ISCC'99*, Sharm El Sheikh, Egypt, July 1999.

[20] M. Mohorcic, A. Svigelj, G. Kandus and M. Werner, 'Performance Evaluation of Adaptive Routing Algorithm in ISL Network', *Networking 2000, Mini-Conference Broadband Satellite Networking*, Paris, France, May 2000.

[21] M. Mohorcic, A. Svigelj, G. Kandus and M. Werner, 'Performance evaluation of adaptive routing algorithms in packet switched intersatellite link networks', *International Journal of Satellite Communications*, submitted for publication, March 2000.

[22] M. Mohorcic, M. Werner, A. Svigelj and G. Kandus, 'Alternate link routing for traffic engineering in pack-etoriented ISL networks', *International Journal of Satellite Communications*, submitted for publication, June 2000.

[23] T. Örs, M. Annoni and V. Paxal, 'A Network Architecture to Interconnect Local Area Networks using non-GEO Satellites', *COST252, 253 and 255 Workshop*, Toulouse, France, May 1999.

[24] E. Papapetrou, I. Gragopoulos and F.-N. Pavlidou, 'Performance evaluation of LEO satellite constellations with inter-satellite links under self-similar and Poisson traffic', *International Journal of Satellite Communications* Vol. 17, pp. 51–64, 1999.

[25] E. Papapetrou and F.-N. Pavlidou, 'Various Routing Techniques for non-GEO Satellite Constellations', *ICT 2000*, Acapulco, Mexico, in press, May 2000.

[26] E. Papapetrou and F.-N. Pavlidou, 'Performance of shortest path routing under various link cost metrics for non-GEO satellite systems', *PIMRC 2000*, London, UK, in press, September 2000.

[27] F.N. Pavlidou, M. Annoni, J. Aracil, H. Cruickshank, L. Franck, T. Ors and E. Papapetrou, 'Traffic Character-ization, Routing and Security Issues in High Speed Networks Interconnected through LEO Constellations', *Joint International Workshop*, Supaero, Toulouse, France, May 1999.

Appendix D

COST253 Members

Belgium	Prof. G. Latouche	Universite Libre de Bruxelles
France	Prof. G. Maral	Ecole Nationale Superieure des Telecommunications-Site de Toulouse
Germany	Mr. U. Freund	Technische Universitat Llmenau
Greece	Prof. N. Pavlidou	National Technical University of Athens
	Prof. P. Constantinou	Aristotle University of Thessaloniki
Ireland	Prof. F.G. Foster	Trinity College Dublin
Italy	Dr. E. Ferro	CNUCE/C.N.R.
	Mr. M. Annoni	CSELT
Norway	Dr. V. Paxal	Telenor
	Mr. A. Nordbotten	Telenor
Slovenia	Prof. G. Kandus	Institut Jozef Stefan
	Dr. T. Javornik	Institut Jozef Stefan
Spain	Dr. J. Gavilan	University of Cantabria
	Prof. J. Garcia	University of Cantabria
UK	Prof. B. Evans	University of Surrey
	Dr. Y. Fun Hu	University of Bradford

Index

AAL, 21, 142, 143
Acquisition, 64
Adaptive routing, 94, 99, 118, 120, 121, 127
Additive White Guassian Noise (AWGN), 39, 40, 49, 58, 63
ad hoc protocols, 212
Adjacent channel interference, 39
Amplitude modulation, 38, 39
AMS, 155
ATM, 1, 3–5, 16–25, 27, 54, 80–82, 86, 88, 89, 91, 110, 114, 115, 128, 129, 131, 132, 139, 141, 142, 146, 148, 225, 226, 217, 218, 230
  PRM, 83, 142, 143
  Service, 19
  ATM-AM, 10, 12, 83
  Security, 142, 143, 145
AristoteLEO, 183–186
Authentication, 140, 141, 143–147
Autocorrelation, 22
Automatic gain control, 39
Autonomous system, 211
Autoregressive model, 22
Autoregressive Moving Average Model (ARMA), 23
Available Bit Rate (ABR), 20, 125
Availability, 80, 83, 84, 96, 118, 120, 132, 134, 138, 148

Bandwidth delay product, 199, 200, 203

BER, 38–42, 45–49, 57, 64–70, 75–77
Binary energy, 49
Binding updates, 211
Binomial distribution, 29
Birth-death process, 29
Bit rate distribution, 22
Block Coded Modulation (BCM), 52, 53
BONeS, 154
Boarder Gateway Protocol, 211, 212
Bose–Chaudhuri–Hocquenghem code (BCH), 50, 54–56
Broadcast period, 121, 122
Bulk data transmission, 24
Burstiness, 18, 19, 22, 25–27, 32, 91

Call admission control (CAC), 16, 93, 125, 156
Care-of-Address (CoA), 210
Cell Delay Variation (CDV), 16–18, 20, 21, 35
Cell Delay Variation Tolerance (CDVT), 16, 17, 20, 35
Cell Error Ratio (CER), 18, 21
Cell Loss Priority (CLP), 126
Cell loss ratio, 18, 20, 21, 125
Cell misinsertion ratio, 18, 21
Cell switching, 86–88
Cell transfer ratio, 18, 35
Celestri, 80, 96, 97, 101, 102, 106–108, 110, 128, 129, 132, 148
Centralised routing, 124

Certification authorities, 146
Chase algorithm, 55
Cipher Block Chaining (CBC), 146
Circuit switch, 86–88
Code Division Multiple Access (CDMA),
    38, 70, 77
Coefficient of variation, 22
Common resource tool, 92
Concatenated codes, 49, 51, 52
    Parallel, 53
    Serial, 54
Congestion avoidance, 35, 196–198, 204
Congestion window, 196, 205
Co-channel interference, 70
Conformance definition, 16
Congestion control, 20, 25, 26, 36, 79, 88,
    125, 126
    Predictive, 125
    Reactive, 125
Connection admission control *see also* call
    admission control, 16, 93
Connection traffic descriptor, 16
Constant Bit Rate (CBR), 19–21, 35, 92, 93
Continuous Phase Modulation (CPM),
    37–39, 41, 42, 76
Convolution codes, 49–53, 59
CONSIM™, 135, 154, 178–183
Core Base Tree (CBT), 130
Correlation, 22–24, 31
    Noise, 62
    Cross, 63
    Line, 22
    Frame, 22
    Scene, 22
CPFSK, 44

DACK, 196
Darpa packet radio routing, 117
Data Encryption Standard (DES), 141,
    144–146
Decentralised routing, 124
Decryption, 142, 145
Dedicated Control Channel (DCCH), 141
Dense mode multicast, 130
Deterministic transformation, 32
Differential pulse code modulation, 21

Differential QPSK (DQPSK), 39, 40
Digital signature, 141, 146
Dijkstra path algorithm, 95, 100, 101, 110,
    111, 113–116
Distance Vector Multicast Routing Protocol
    (DVMRP), 130
Distributed routing, 94, 124
Distribution of scene duration, 22
Döppler shift, 57–69, 76–78
Double coverage area, 70
Double moveable boundary strategy, 92
Dynamic Host Configuration Protocol
    (DHCP), 210, 212

Equivalent terminal, 16
Encapsulation, 144
Encryption, 140–142, 144–146
Euclidean distance, 41, 46, 52, 54, 55, 78
Euroskyway, 3, 4
Explicit rate, 126
Exterior Gateway Protocol (EGP), 211
Eye closure, 44–47, 75

Fast-Packet switching, 86–88
Fast recovery, 196–198
Fast retransmission, 196–198, 206
Feedback loop, 57, 59, 64, 65, 67, 69, 76–78
Flow Deviation Algorithm (FD), 110, 111,
    113–116, 147
Foreign agent, 209–212
Fractional Brownian Motion, 32–33
Fractal Renew process, 32

Galois Field, 50
GaliLEO, 154, 162–178
Gaussian-shaped GMSK, 41
Generalised Bass model, 33
Generic Cell Rate Algorithm (GCRA), 16,
    17
Geographic traffic distribution, 33, 79, 95,
    98
Geostationary Satellites, 5, 139, 146–148
Geometrically Modulated Deterministic
    Process (GMDP), 24, 26, 27, 35
Globalstar, 216, 224
Gray mapping, 54

Gross domestic product, 98
GSM, 37, 132, 140, 141, 148
Guaranteed frame rate, 20, 21

Hamming code, 50
Hamming distance, 52, 54–56
Handover, 79, 91, 118, 124, 146, 157
High-Altitude Aeronautical Platform
    (HAAP), 38
High Power Amplifier, 39, 41, 42
Home agent, 209–212
HTTP, 196
Hurst parameter, 32
Hybrid switching, 86–88

Interrupted Bernoulli process, 29
Interrupted Poisson Process, 21, 24, 28
Interactive data transmission, 24
Intermediate system to Intermediate system
    routing (IS-IS), 117
Internet Engineering Task Force (IETF), 86,
    87, 144, 195, 196
Internet Group Management Protocol
    (IGMP), 130
Inter-Satellite Link (ISL), 6, 80, 85, 91,
    95–101, 103–105, 107–110, 113, 115,
    117–122, 126, 127, 129, 131, 138,
    156–159, 163, 166, 168, 169, 172–175,
    178, 184, 189, 190, 218, 219, 221–223,
    225, 226
Intersymbol interference, 60
Intrinsic Burst Tolerance (IBT), 16, 17, 19
Iridium, 2, 3, 79, 118–121, 123, 124, 216,
    225,
Intermittent Connectivity, 199, 205, 212
Internet Protocol (IP), 80–83, 85, 86,
    130–132, 139, 144, 148, 195, 194–212
Ipv4, 210
Ipv6, 26, 210, 211
IP Adapter Module (IP-AM), 83
IP encapsulation, 211
IPSEC, 144, 211
ISO Inter-Domain Routing Protocol
    (IDRP), 211
Isolated routing, 121

Isotropic Phase Type Planar Point Process
    (IPHP3), 38, 70–72, 77

Jitter, 57, 64, 65, 67–70, 77

LAN ATM converter, 82
Legacy routing, 95
LeoSim, 155, 161
Level switching, 44
Linear block codes, 50
Link cost, 99, 100, 101, 104, 105, 107
Logical link control, 81

MAC, 81, 147
MAMSK, 43
Markov Modulated Poisson Process
    (MMPP), 21, 23, 24, 28, 31, 32
Maximum Burst Size (MBS), 16, 17, 20
Maximum Likelihood Sequence Estimation
    (MLSE), 42, 46, 50
Mean burst duration, 18
Minimum cell rate, 20
Mobile IP, 209–212
MPSK, 41
M-Star, 4, 118–120, 123, 128,
Multicast, 127, 130–132, 148
Multicast Open Shortest Path First
    (MOSPF), 130, 131
Multistate Markov source model, 26
Mutual authentication codes, 147

N-Minimum Shift Keying (N-MSK), 38,
    42–44, 47, 48, 75
Non-uniform activity-level, 23
Noise power spectral density, 49
Nyquist filter, 39, 40, 58–60, 63, 65, 68, 76
Odyssey, 96
On-board routing, 6
On-board processing, 84, 86, 87, 90, 133,
    136
On-board switching, 80, 85, 87, 225, 226,
    228, 230
On-demand route, 123, 124
On-off model, 27
Open Shortest Path First (OSPF), 117

Operation and Maintenance (O&M), 84,
    133–135
OPNET, 154
Orbits, 5, 80, 83, 85, 95, 96, 105, 107, 108,
    119, 131, 133, 137
    Circular, 215, 216, 222, 228
    Elliptical, 215–218
    Geostationary Orbit (GEO), 5, 216, 217,
        222, 226–229
    Low Earth Orbit (LEO), 5, 216, 225, 226,
        228, 229
    Medium Earth Orbit (MEO), 6, 216, 225,
        227–229
OSI, 25, 158

Packet switching, 86, 87, 90
Parallel Concatenated Convolutional codes
    (PCC), 53, 54
Pareto distribution, 33, 109
Path delay, 80, 85
Path loss, 80, 83, 85
Path computation, 94
Path metrics, 94
Path selection, 94
PAWS, 200
Peak Cell Rate (PCR), 16, 20, 35
Peak emission interval, 17, 35
Period scheme, 121
Poisson, 21, 24, 29, 31–33, 99, 112, 115
POTS, 21
Pre-computed routes, 123
Propagation delay, 79, 95, 100, 101,
    103–107, 118, 120, 125, 126, 132
Protocol Independent Multicast Sparse
    Mode (PIM-SM), 131
Private Network to Network Interface
    routing (PNNI), 117

QoS, 15, 18, 81, 85, 86, 95, 124, 125,
    132–134, 140, 153, 156, 159, 161, 163,
    168, 170–173, 192
Quadrature Double DPSK (QDDPSK), 63,
    68, 69, 77
Quadrature Differential Phase Shift-Keying
    (QDPSK), 60, 61, 63, 65–68, 76, 77

Quaternary Phase Shift Keying (QPSK), 39,
    40, 42, 54, 55, 57, 59, 60, 62, 64, 65
Quadrature Amplitude Modulation (QAM),
    39–42, 44, 52, 53, 64, 75, 76, 78

Rain fading, 5
Raise Cosine Pulse Length (LRC), 41
Received Signal Strength Indicator (RSSI),
    44
Rectangular Frequency Pulse Length
    (LREC), 41
Reed-Solomon code, 50, 51
Regenerative transponders, 86
Reliability, 79, 80, 84, 87, 132–134,
    136–139, 148
Reliability block diagram, 134, 136
Reply factor, 100
Resource control, 79, 91
Resource management, 126, 143
RESQ, 154
Retransmission time out, 200, 203, 209
Route diversity, 118–120
Routing, 94, 96, 118, 120, 121, 123–125,
    127, 130, 131, 136, 138, 142, 146–148,
    205, 210, 211
Routing Information Protocol (RIP), 117
RSA, 141

SACK, 206, 212
Satellite coverage, 95
Satellite Personal Communications Services
    (S-PCS), 1
Satellite Switch Time Division Multiple
    Access (SSTDMA), 6
Satellite Network Termination Modules
    (SNTM), 80, 82
SEESAWS, 186–192
Security, 139–144, 146–148
    User plane, 143
    Control plane, 143
    Management plane, 143
    Association, 144
Self-similar traffic, 32, 33, 100, 108, 109,
    114, 116, 117, 147
Severed Error Cell Block Ratio (SECBR),
    18

Shannon Theorem, 49
Shannon limits, 53, 76
Signal-to-Interference Ratio (SIR), 71
Slow start threshold (ssthresh), 196–198,
Sparse mode multicast protocol, 130
Skybridge, 3, 4, 127, 132, 215, 216, 221,
    222, 226, 227
Source model, 26
Source traffic descriptor, 16
Spectral Power Density (SPD), 59, 63, 68,
    69
Spectral Raised Cosine (SRC), 41
Static routing, 120
Statistical multiplexing, 81, 87
Subscriber Identity Module (SIM), 140
Sustainable Cell Rate (SCR), 16–20
Switched Poisson Process (SPP), 24, 31
Synchronous Digital Hierarchy (SDH), 83
Synchronous transfer mode, 19

Telecommunications Industries of America
    (TIA), 80
Teledesic, 3, 127, 132, 215, 216, 221, 222,
    224–226
Terrestrial network link routing, 117
Terrestrial Network Termination Modules
    (TNTM), 80, 82
Tamed Frequency Modulation (TFM), 41
Time Division Multiple Access (TDMA),
    92, 93
Time stamp, 212

Time-to-live, 131
Traffic Control Channel (TCCH), 141
Traffic flow distribution, 98, 107
Traffic intensity, 33, 97–99, 123
Traffic models, 15, 16, 24, 33
Traffic weight factor, 100
Transmission Control Protocol (TCP), 20,
    25, 33, 195–212
Trellis Coded Modulation (TCM), 52
Triangular routing, 210
Trigger threshold, 122
Tunnelling, 82
Turbo code, 38, 52, 54, 75, 76, 131
Type of Service, 95

Uniform activity-level scene, 22
Uniform traffic, 100, 101, 104
Unspecified Bit Rate (UBR), 20

Variable Bit Rate (VBR), 17, 19
Variable rate N-MSK, 42, 75
Viterbi, 42, 46, 50, 51
Virtual path, 110, 131, 142, 143
Virtual path connection, 110, 143
Virtual circuit, 110, 143
Virtual destination, 126
Virtual source, 126
Voltage Controlled Oscillation (VCO), 59

WEST, 3, 4, 80, 215, 216, 227, 228,
White noise, 22

Printed and bound by CPI Group (UK) Ltd, Croydon, CR0 4YY

Printed and bound by CPI Group (UK) Ltd, Croydon, CR0 4YY

27/10/2024

14580219-0005